Cosmological Frontiers

By Dr. Anab Whitehouse

© Anab Whitehouse
The Interrogative Imperative Institute
Brewer, Maine
04412

All rights are reserved. With the exception of uses that are in compliance with the 'Fair Usage' clause of the Copyright Act, no portion of this publication may be reproduced in any form without the express written permission of the publisher. Furthermore, no part of this book may be stored in a retrieval system, nor transmitted in any form or by any means - whether electronic, mechanical, photo - reproduction or otherwise - without authorization from the publisher or unless purchased from the publisher or a designated agent in such a format.

Published by: Bilquess Press
2018

The greatest obstacle to discovery is not ignorance – it is the illusion of knowledge.

Daniel J. Boorstin

## Table of Contents

Foreword – page 7

Chapter 1: Meet the Champ – page 11

Chapter 2: The Meaning of Red – page 33

Chapter 3: Noise – page 51

Chapter 4: The Electric Universe – page 85

Chapter 5: Matters of Gravity – page 109

Chapter 6: Mysterious Holes – page 145

Chapter 7: Through A Glass Darkly – page 175

Chapter 8: Expanding Horizons – page 207

Chapter 9: Branes for Hire – page 239

Chapter 10: Odds and an End – page 277

Bibliography – page 299

| Cosmological Frontiers |

6

### Foreword

While many people today – especially scientists and the people over whom the former individuals have influential sway – are under the impression that science offers the best way to discover the nature of reality, nonetheless, an array of considerations have been put forward in my other written works -- including the current volume -- which seek to give credence to the possibility that however valuable science might be as a means of engaging a variety of physical problems, nevertheless, epistemologically speaking, science still leaves much to be desired, and, consequently, many unknowns and uncertainties permeate the fabric of the sciences.

I am not convinced that science – despite its heuristic value -- is the best way to go about trying to resolve the reality problem in a temporal context that is severely limited by the demands (e.g., work, sleep, family, community, school, physical needs) placed on the uses to which such time is put. One is likely to be dead and gone for quite some time before the methodologies of science and mathematics will be able to make even limited headway concerning the nature of physical reality ... not to mention trying to make progress with respect to issues involving the nature of consciousness, intelligence, creativity, talent, reason, language, morality, and spirituality, and, yet, one is faced with the problem of having to deal with life and make decisions about how to proceed despite being immersed in many unknowns.

More than half a century ago, C.P. Snow introduced the notion of "Two Cultures" (i.e., humanities and sciences) to talk about different ways of engaging the reality problem. In many respects, that notion might be far too limiting, and, consequently, we should not necessarily restrict ourselves to what the foregoing two kinds of disciplines have to offer in the way of ideas methodologies, or theories, and, instead, we need to become focused on critically engaging those disciplines with the purpose of trying to discover the nature of truth quite independently of whether what is found has the stamp of approval of either the humanities or the sciences.

There is, I believe, a third culture (a culture of truth) that transcends, even as it includes, elements of science and humanities and depends on something more than the methodologies inherent in those two disciplines. The process of critical reflection that is at the heart of the pages of this book is intended to give expression to some considerations beyond science and humanities that might be of value when trying to engage the reality problem with which we all are confronted.

To be sure, such a journey of discovery requires a familiarity with, and understanding of, the work being done in both the humanities and sciences. That is, one must have a certain level of literacy when it comes to understanding and appreciating what the humanities and sciences have to offer.

Nonetheless, the hybrid epistemological/hermeneutical vehicle by means of which one makes ones way through life must be capable of running on a fuel that transcends both of the foregoing disciplines. This fuel must have the potential to attain conceptual escape velocity and not just putter along with an orbital velocity that is buffeted about by the social, institutional, ideological, and historical forces to which the practice of humanities and sciences often tend to give expression.

Of course, when one begins to experiment with the composition of fuels – especially ones that are intended to be powerful enough to carry one away from the gravitational pull of such massive bodies as the humanities and the sciences -- there is always the risk that one's efforts will blow up in one's face or end in some other kind of tragedy. Knowing what one needs to do in life and being able to realize that intention are not necessarily synonymous with one another.

Throughout my written works, I have been attempting to follow something akin to the method attributed to Michelangelo (the possibility that this attribution might be apocryphal is irrelevant). In other words, I have been seeking to chisel away, or remove, whatever elements seem not to belong as far as my conceptual sculpture of "The Reality Problem" is concerned.

There is an interstitial quality to the foregoing chiseling process. The conceptual or hermeneutical sculpture to which I am alluding is not so much a function of whatever substantive facts remain after the chipping activity has been completed, as much as the intended figure or object of understanding to which attention is being drawn tends to reside in the conceptual spaces beyond and between those factual residues, just as the placing of two appropriately shaped vases (or candlesticks) creates the image of several facial profiles in the space between those vases (candlesticks).

The chiseling process is critical reflection. Critical reflection is not just a function of reasoning of one kind or another.

Critical reflection gives expression to everything within us – experience, needs, interests, intelligence, rationality, emotions, intuition, imagination, the 'self', creativity, curiosity, questions, judgments, and so on – that is intent on trying to find the truths inherent in reality. Critical reflection is a reiterative process that continues to feed the results of previous rounds of critical reflection through the grinding process which constitutes critical reflection – that is, the constant process of: (1) Asking questions concerning, (2) posing problems with respect to, (3) rigorously examining the properties of, (4) probing the possibilities inherent in, and (5) evaluating the strengths and weakness entailed by the data of experience ... both mine and that of others.

Critical reflection is the fuel that help makes possible (but might not be solely responsible for) the achievement of escape velocity possible with respect to the gravitational pull of the humanities and sciences. If one makes mistakes with respect to the composition, refinement, and use of the sorts of fuels being alluded to, one's attempted journey to the realms of truth that lie within, as well as beyond, the humanities and sciences is likely to suffer delays, setbacks and problems ... if not disaster.

Everyone is responsible for his or her own fuel work and the consequences that ensue from such work. No one has the right to impose his or her solutions with respect to that sort of work on other human beings, and, in addition, one bears responsibility for whatever difficulties that work causes in relation to other human beings.

On the other hand, sharing potential solutions with others, or consulting with one another concerning those possibilities, seems eminently reasonable and, potentially, quite constructive. This is the spirit with which this book being written.

| Cosmological Frontiers |

| Cosmological Frontiers |

Chapter 1: Meet the Champ

When Einstein released his ideas on General Relativity in 1915, the dominant model of the universe was a variant on what became known as the 'Static or Steady State Theory of the Universe'. Although the latter model has assumed a variety of forms over the years, the basic idea was that the universe has always existed, and the manner in which the cosmos operated could be described through the laws of physics ... such as those that were given expression through general relativity (i.e., Einstein's re-visioning of Newton's theory of gravity).

In 1917, Einstein introduced a cosmological constant – lambda, $\Lambda$ -- into his earlier field equations to account for why the universe did not collapse under the constant pull of universal gravitational attraction. The foregoing constant alluded to the presence of some sort of force that resisted the presence of gravitational attraction by an amount that was sufficient to keep things pretty much on a "steady-as-she-goes" heading.

The addition of the aforementioned cosmological constant was later referred to by Einstein as being his greatest blunder. Supposedly, the nature of the error was laid bare through the work of, among others, Edwin Hubble in the 1920s that was rooted in empirical data indicating that the universe might be expanding.

Even before Hubble undertook his groundbreaking work in astronomy, a Russian mathematician, Alexander Friedman, had given Einstein's field equations a workout and appeared to show there were solutions to those equations indicating that the universe was expanding. Einstein disagreed with Friedman's conclusions because Einstein was committed to the idea of a -- relatively speaking -- static universe, but, apparently, there were hidden dimensions in the equations of general relativity that even Einstein had not suspected (a scenario that would be played out again in relation to the issue of "black holes").

Einstein's alleged "blunder" would become rehabilitated – possibly --more than half a century later when astronomical data seemed to demonstrate that the universe was expanding as a result of the presence of a force referred to by many as "dark energy". Einstein's notion of a cosmological constant appeared to be intimately connected with the – purportedly -- newly discovered force, and, therefore,

Einstein's alleged blunder might actually turn out to be a prescient intuition concerning the nature of a very significant dimension of the universe (There will be more discussion on this topic later on in the book.).

Despite the pronouncements of individuals such as Immanuel Kant that gave expression to the idea that the universe consisted of many galaxies, nonetheless, until the 1920s, most scientists believed the Milky Way was the only galaxy in the universe. Indeed, such individuals considered the Milky Way and the universe to be, more or less, coextensive.

The nebulae that could be observed through telescopes were interpreted as clouds of dust and gas that, eventually, might coalesce into stars. However, such clouds – along with their possible, subsequent development into stars -- were considered to be phenomena that took place fully within the Milky Way galaxy.

One of the foregoing nebulae was known as M31. In 1924, using the 100-inch telescope located at the summit -- a little over a mile high -- of Mt. Wilson (near Pasadena, California), Edwin Hubble undertook the task of trying to measure the distance to M31. Hubble's method was based on the absolute and apparent luminosity properties of stars known as Cepheid variables ... properties that could be quantified and, in turn, be fed into a distance formula for determining, within limits (up to about 163 million light years), how far away a given cosmic object might be.

Hubble's calculations were off by a factor of 2. That is, he calculated the distance to M31 to be twice as close as it actually was ... a mistake that was corrected in the 1950s.

Nevertheless, despite the error in calculation, the distance to M31 (which is now known to be 2.5 million light years away) was much, much farther from Earth than the most distant known stars that existed in the Milky Way. M31 seemed to exist in a realm beyond the confines of the Milky Way galaxy and, subsequently, came to be known as the Andromeda galaxy.

Hubble measured the distance to a number of other nebulae that could be viewed through the telescope at Mt. Wilson. Some of them were calculated to be hundreds of thousands of times more distant than the most distant stars in the Milky Way galaxy.

The Milky Way galaxy was not the only inhabitant of the universe. It could no longer be considered to encompass the sum total of physical reality.

Hubble also noticed another relationship between the properties of the cosmic objects he was studying and their distance from Earth. If one examined the emission or absorption lines in the spectra of the light given off by those heavenly bodies, there appeared to be an inverse relationship involving luminosity and distance.

More specifically, the dimness of those objects tended to be correlated with measurements indicating that such objects were quite distant from Earth. That is, the greater the degree of dimness, the more distant those objects were measured to be.

Dimness also seemed to be related to a shift in the frequency of the spectral lines associated with such sources. This transition toward a lower frequency of the spectrum is known as a redshift.

However, the spectral properties of cosmic objects did not always involve shifts toward the red end of the spectrum. For instance, M31, or the Andromeda galaxy, exhibited a blue shift (i.e., toward the blue end of the spectrum) in the absorption/emission properties of its light.

Nonetheless, by 1925, the predominant tendency displayed in the spectral properties of cosmic object studied by Hubble involved redshifts, rather than blueshifts. Four years later, in 1929, Hubble proposed a law that governed the relationship between distance and redshift.

The law-like relationship was linear in nature. This meant that the dimmer a cosmic object appeared to be, then the more the spectral properties displayed by such objects tended to be shifted in the direction of the red end of the spectrum. As a result, redshifts became associated with the idea of distance.

Furthermore, spectral redshifts also became associated with the notion of recessional velocity ... that is, the speed with which some given cosmic object appears to be receding from Earth. The greater the degree of redshift, then, the greater the recessional velocity of that object relative to Earth was considered to be.

The idea of recessional velocity was tied to the Doppler effect. In other words, just as sound waves exhibit higher and lower frequencies as they travel, respectively, toward us and away from us, so too, the shift toward the red end of the spectrum that was exhibited by the spectral properties of light coming from cosmic objects was interpreted to mean that the source emanating such light was moving away from Earth, just as a shift toward the blue end of the spectrum was interpreted to mean that the source generating such light was moving toward the Earth.

Notwithstanding the fact that the very first cosmic object studied by Hubble – namely, M31 or the Andromeda galaxy – exhibited a blueshift in its spectral properties, astronomers began to interpret Hubble's data as indicating that the universe was expanding. This seemed to validate Friedman's earlier understanding of Einstein's field equations in which the former individual (i.e., Freidman) believed that Einstein's equations indicated that the universe was expanding despite Einstein's resistance to such a possibility.

Yet, if the universe was expanding, then, why – according to its spectral properties – did M31 seem to be speeding toward Earth? Although most of the cosmic objects studied by Hubble displayed a redshift in its spectral properties, this was not always the case, and, so, what was one to make of a universe that contained objects that did not seem to be caught up in the general move away from Earth?

How were spectral redshifts and blueshifts to be reconciled with one another? Or, stated in a slightly different way, what was the significance of spectral blueshifts in a universe that seemed to be dominated by cosmic objects displaying spectral redshifts?

The network of interconnecting relationships underlying the notion of an expanding universe appeared to consist of: Dimness, distance, redshift, and recessional velocity. Yet, maybe, in some instances, dimness was not necessarily a marker for distance but, instead, indicated one was dealing with something that was merely dim and, for whatever reason, either giving off limited luminosity or displaying a form of luminosity that was, in some way, filtered during its journey to Earth.

Alternatively, maybe redshifts didn't necessarily always indicate that the sources of such spectral wavelength shifts were moving away from Earth. Maybe there were other, possible interpretations for the significance of redshifts ... an issue that will be explored in Chapter 2.

Let's assume for the moment, however, that the dimness of cosmic objects is a sign of distance and that redshifts signify recessional velocity. Given such assumptions, how did things proceed within astronomy from that point?

Enter Georges Lemaître. Before becoming a professor of astrophysics, Lemaître was a Roman Catholic priest.

Lemaître was interested in science. Moreover, following in the footsteps of many natural philosophers that preceded him, Lemaître saw nature as an active function of God's presence and, therefore, the pursuit of science was something that he believed was eminently reconcilable with his spiritual perspective.

His scientific training was rigorous. He received his doctorate from MIT.

Prior to receiving his doctorate, he worked in England with Arthur Eddington who initiated Lemaître into the disciplines of astronomy and cosmology. Lemaître followed up on his University of Cambridge studies by working with Harlow Shapley, a well-known astronomer, at the Harvard College Observatory.

Lemaître returned to Belgium in 1925. He became a part-time lecturer at the Catholic Universe of Leuven.

In 1927 he wrote an article that appeared in the *Annals of the Scientific Society of Brussels*. The paper gave expression to a theory concerning the idea of an expanding universe.

Edwin Hubble -- who is often cited as the first scientist to propose the idea of an expanding universe in 1929 -- was beaten to the punch by Lemaître's 1927 paper. However, as far as the notion of an expanding universe is concerned, perhaps, ultimate priority should be given to the previously mentioned Alexander Friedman who, in 1922, had derived solutions from Einstein's general relativity field equations indicating that the universe was expanding.

Issues of priority aside, Lemaître's foregoing article was written for a publication that did not receive much attention in the world of astronomy beyond the borders of Belgium. Consequently, his ideas about an expanding universe went largely unnoticed.

Four years later in 1931, Lemaître – with assistance from his former mentor, Arthur Eddington – translated his 1927 article into English. Einstein became aware of Lemaître's ideas concerning an expanding universe and indicated to Lemaître that while the latter's calculations were acceptable Lemaître's physics (i.e., the notion of an expanding universe) were "atrocious".

As noted previously in this chapter, Einstein had rejected Friedman's similar ideas involving the notion of an expanding universe nearly a decade earlier. Apparently, Einstein saw nothing in the work of Lemaître that changed his mind with respect to the tenability of a static or steady universe.

Friedman's work was largely mathematical in nature. Einstein did not accept the former individual's solutions to the field equations of general relativity.

Lemaître's treatment of the expanding universe idea involved more than a mathematical reworking of Einstein's field equations. It was a theory in astronomy that attempted to make sense of, among other things, the behavior of nebulae, and Eddington felt that Lemaître's idea concerning an expanding universe resolved a number of problems in cosmology.

After Lemaître's ideas were translated into English, he was invited to speak on them in London. Lemaître took that opportunity to introduce the notion that the universe had expanded from some initial point that he referred to as a "Primeval Atom".

Later on, Lemaître described his "Primeval Atom" as a sort of "Cosmic Egg" that began to explosively unpack its potential at the moment of Creation. As far as Lemaître was concerned, such terms were just alternative ways of giving expression to the idea of an expanding universe.

Although Eddington initially had supported Lemaître's idea of an expanding universe, he was less enthusiastic about the notion of a "Primeval Atom" or "Cosmic Egg" that gave rise to an expanding universe that exploded onto the scene at the moment of Creation. Einstein, on the other hand, believed that Lemaître's physics were wrong and, consequently, that the idea of an expanding universe could not be demonstrated.

Fred Hoyle was a respected British astronomer. In the 1940s, Hoyle developed -- along with Thomas Gold, and Hermann Bondi -- a steady state theory of the universe. Among other things, the theory being alluded to posited that the universe had no beginning and no end.

During an episode of Hoyle's BBC radio program, *The Nature of the Universe*, he critiqued Lemaître-like theories. At one point in the program, he used the term "Big Bang" to dismissively refer to such ideas.

While Lemaître and Hubble were developing their respective theories concerning the idea of an expanding universe, Alexander Friedman was continuing to develop and disseminate his own ideas in Russia with respect to the notion of an expanding universe. Among his students was a brilliant individual, Georgy Gamov.

Friedman died in 1925 from typhus ... just three years after deriving his solutions to Einstein's field equations. Nevertheless, Friedman still managed to spend considerable time with Gamov, initiating the latter individual into, among other things, Friedman's cosmological take on Einstein's theory of gravity.

In 1934, Gamov moved to the United States and became known as George Gamow. Subsequently, he accepted a faculty position at George Washington University.

While Gamow explored a variety of areas in science – including radioactive decay, the formation of stars, and nucleosynthesis (the generation of atoms that are more complex than hydrogen) – he also was an advocate for, and contributor to the development of, the theory of Big Bang cosmology. Gamow re-envisioned Lemaître's Creation-based, expanding universe and presented those ideas in purely physical terms (that is, without any mention of Creation or a Creator).

Gamow believed that in the early universe, radiation predominated over matter. As a result, things were hot.

Using quantum mechanics, general relativity theory, and a variety of other discoveries of 20th century physics, Gamow worked out a temporal sequence in which the universe proceeded from a hot, radiation-dominated realm through to, over time, the development of stars and galaxies in an expanding universe. In addition, Gamow advanced theories about how -- during the aforementioned period of expanding development -- different atomic elements (hydrogen and beyond) would be produced in the hot, thick particle soup of the Big Bang through a process that is known as nucleosynthesis.

Gamow put a quantitative face on the development of the universe. For example, he calculated at what point the density of matter and radiation would equalize once the Big Bang took place (and, remember, Gamow maintained that in the beginning, radiation dominated over matter).

In addition, Gamow made calculations concerning the density of matter that would be necessary to set the forces of nucleosynthesis in motion. This led, in turn, to theoretical calculations concerning the mass, composition, and size of early galaxies.

As well, Gamow produced several quantitative predictions for the temperature of the radiation that would remain in the background as remnants of the initial Big Bang and the subsequent early expansion of the universe. This was the first prediction concerning the temperature value for what, today, is referred to as Cosmic Background Radiation.

Not everything that Gamow calculated and theorized has stood the test of time. Nevertheless, his work – along with the contributions of individuals such as Ralph Alpher (a former graduate student of Gamow's), Robert Herman, and Hans Bethe – shaped much of the staging area from which ensuing theories of Big Bang cosmology have been launched.

Up until the time of Gamow's work -- and despite the contributions of individuals such as Alexander Freidman, Georges Lemaître, and Edwin Hubble -- many scientists still seemed to be inclined toward the static or steady state-like universe of Albert Einstein and Fred Hoyle. After Gamow introduced his ideas, an increasing number of scientists began to move in the direction of a Big Bang theory of some kind.

Perhaps one of the reasons for the foregoing shift in beliefs toward Big Bang cosmologies and away from Steady State cosmologies had to do with what Gamow' work provided that no other astronomer prior to him had been able to offer ... except in limited ways. More specifically, Gamow had put forth a plausible narrative concerning how the universe might have made the transition from: An early hot, radiation-dominated set of conditions, to: A universe dominated by matter, gravity, and the accretion of materials that led to the formation of stars and galaxies.

Gamow's ideas reflected, and were rooted in, the work of, among others, Isaac Newton, Max Planck, Albert Einstein, Arnold Sommerfeld (fine structure constant), Alexander Friedman, Georges Lemaître, and Edwin Hubble. Yet, at the same time, Gamow pointed to possibilities that both united and extended the earlier work within the context of a coherent, scientific narrative that explained – within certain limits -- how such a cosmology was consistent with, and might account for, a great deal of empirical data.

Eight years after Gamow passed away in 1968, Steven Weinberg wrote a book entitled: *The First Three Minutes: A Modern View of the Origin of the Universe*. The book was an expanded version of a talk that Weinberg had given in 1973 in conjunction with the dedication of Harvard's Undergraduate Science Center.

The book was written six years before Weinberg received a Nobel Prize in physics – along with Abdus Salam and Sheldon Glashow – for work on the electro-weak theory of quantum dynamics. Therefore, when the foregoing book was published, Weinberg was not a household name, but he was still a first-rate physicist.

Weinberg professional interests were mostly directed toward particle physicist. He was not an astronomer or cosmologist.

Nonetheless, particle physics played a substantial role in Big Bang Cosmology. Consequently, Weinberg used his expertise in the former area to deliver a relatively popularized treatment of the latter topic that is still considered by many individuals to be relevant nearly forty years later.

*The First Three Minutes* took off where Gamow left off. Weinberg's book was an updated and expanded version of the Big Bang cosmology that had been developed by Gamow, and others, through the 1950s and 1960s.

During the Introduction to his book, Weinberg spoke about the initial explosion that marked the advent of the universe. He described that event as unlike the "normal" sort of explosion that emanates outward from a determinate center.

Instead, the Big Bang supposedly happened simultaneously everywhere in space. As this occurred, each particle was sent flying away from every other particle.

In addition, Weinberg notes that such an omnipresent "explosion" could have taken place in space that was infinite in nature or might have taken place in space that curved back on itself and, therefore, was finite. He did not feel the nature of space – that is, whether it was infinite or finite – really affected what transpired during the first three minutes.

Since the subtitle to his book is: "A Modern View of the Origin of the Universe", one might note in passing that such a description is somewhat misleading. For example, in the aforementioned introductory remarks, he indicates that the space in which the Big Bang took place already existed, and, as well, he indicates that during the initial explosion, particles were flung away from one another.

Consequently, both space (whether finite or infinite in nature) and particles existed prior to the Big Bang. Moreover, some mechanism or force or form of energy must have existed that resulted in an explosion that occurred everywhere in space.

Weinberg's book does not explain the origins of space, particles, or the capacity that underwrote a universal explosion. Instead, he assumes the existence of such things and proceeds forward from that presumptive starting point in order to try to account for how the universe unfolded after the aforementioned initial explosion.

One also should note that Weinberg's starting point is quite different from that of Georges Lemaître. The latter individual indicated that some sort of Primeval Atom or Cosmic Egg existed prior to the Big Bang, and it is that 'Atom' or 'Egg' which exploded, whereas Steven Weinberg claims that space exploded everywhere at once, and, as a result, there was no center (i.e., "Atom" or "Egg") involved in such an explosion.

Neither Lemaître nor Weinberg can account for the origins of their respective starting points. Consequently, both versions of the Big Bang are enveloped by various clouds of unknowing.

Although many individuals refer to the Big Bang as if it were a monolithic theory, the fact of the matter is there are at least two editions of that theory. One edition of the theory follows Lemaître -- although the terms: "Primeval Atom" and "Cosmic Egg" have been replaced by the notion of a "singularity" -- while the other version of the Big Bang theory follows George Gamow and speaks in terms of a hot plasma, of some kind, that existed at the beginning of things.

Whether one is talking in terms of singularities or hot plasmas, the nature of that starting point is enveloped in mystery. In addition, the nature of the explosion process is also unknown irrespective of whether one is talking about singularities or hot plasmas.

The term that is used today to allude to that explosion process is: "symmetry breaking". Something happens that pushes the universe out of a state of equilibrium (symmetry condition) and into an event that either rips a singularity apart or causes particles to fly away from one another everywhere in space simultaneously.

According to Steven Weinberg, in the first one-hundredth of a second during which the Big Bang was taking place, the temperature of the universe was somewhere in the vicinity of $10^{11}$ degrees Centigrade ... a temperature that is considered to be far greater than the temperatures believed to exist in the center of the hottest stars. Since Weinberg maintains that space exploded everywhere during the Big Bang, one has a difficult time trying to explain how such elevated temperatures might be possible even in a confined area, let alone everywhere at once ... and, if space is infinite in nature, the foregoing question concerning what made those sorts of extreme temperature possible becomes even more problematic.

The $10^{11}$-Centigrade figure is not an empirical fact. It is a starting assumption ... it is the sort of figure with which one must begin if one hopes to be able to offer a plausible account of what might have happened during the next two minutes, 59 and 99/100-plus seconds.

Weinberg believes that among the many particles that populated the hot plasma existing at the beginning of the Big Bang there were four particles that existed in abundant numbers. These quantum objects were: electrons, positrons, neutrinos, and photons.

The foregoing electrons and positrons being described by Weinberg were considered to have been in near equilibrium with one another. Since in today's world, one observes the presence of positrons only occasionally in relation to certain manifestations of radioactivity or in conjunction with cosmic rays or in relation to high-tech accelerators and colliders, Weinberg's statement about the condition of equilibrium or near-equilibrium between electrons and positrons during the early universe is not an empirical fact but an assumption that might, or might not, be true.

The universe today seems to get along quite well today despite a significant asymmetry between matter and antimatter. In addition, the universe appears to be able to run along quite smoothly with antimatter playing what appears to be a relatively secondary – but, nonetheless, important -- role in a universe that consists largely of matter.

As pointed out in *Volume II* of *Final Jeopardy: Physics and the Reality Problem*, no one currently knows why there is an asymmetry between 'normal' particles and their antimatter counterparts. Consequently, at the present time, no one knows if there were some kind of symmetry existing between electrons and positrons prior to the Big Bang that, in an unknown fashion, became asymmetric during, or following, the Big Bang, or whether, on the other hand, there might have been such an asymmetry from the very beginning.

Of course, Steven Weinberg might be correct with respect to his foregoing assertions. Maybe, the universe really was at a temperature of $10^{11}$ degrees Centigrade in the first 1/100[th] of a second of the Big Bang, and, perhaps, electrons and positrons actually were in equilibrium or near-equilibrium during that first, fraction of a second, but at this scientific point in time, such claims exist because of their narrative value rather than because they give expression to empirical facts.

Weinberg claims that in the early universe electrons, positrons, neutrinos, and photons were continuously: (1) being created out of what he describes as "pure energy", (2) proceeding to live for a short time, and, then, (3) disappearing in the mists of annihilation during collisions of one kind or another. The meaning of the foregoing term "pure energy" is a little unclear ... although it seems to be caught up with the idea of the incredibly high temperatures that supposedly existed in the first $1/100^{th}$ of a second of the Big Bang.

Many people today are familiar with Einstein's idea that there is an equivalency between energy and matter. For instance, under the right sort of conditions, a thermonuclear weapon is capable of transforming a certain amount of enriched uranium material into energy, and, as well, high-tech accelerators and colliders are able to generate levels of energy through the collision of various kinds of accelerated particles with one another, and, during the ensuing aftermath of such collisions, different kinds of particles emerge.

Nonetheless, the transition process from energy to mass is not as well understood as is the transition process from mass to energy. Physicists can calculate what sorts of particles are likely to arise at certain levels of energy, but how energy gets converted into one kind of order rather than another is not known ... that is, scientists do not know how the presence of a certain level of energy coalesces into a particle exhibiting, for example, one combination of electrical charge, spin, mass, etc., rather than some other combination of electrical charge, spin, mass, and so on.

Physicists can predict what sorts of particles are likely to appear given a certain level of energy, and physicists know, as well, that what appears in such a energy context will be in compliance with various laws of conservation and thermodynamic considerations involving stability and instability. Nonetheless, the precise nature of the transition process from energy to matter is shrouded in a certain amount of mystery because no one knows how energy comes to be ordered in one way rather than another.

During the next 14 seconds of the Big Bang, Weinberg indicates that the temperature of the universe dropped. More specifically, after one-tenth of a second, the universe 'cooled' to about $3 \times 10^{10}$ degrees Centigrade, and, continued to drop until at the 14-second marker, the temperature of the universe was around three billion degrees Centigrade.

According to Weinberg, the initial temperature of the Universe was $10^{11}$ degrees Centigrade. In order for such a temperature to cool down, there must have been a means of dissipating the temperatures associated with that energy.

The explosion entailed by the Big Bang supposedly was occurring everywhere in space. Consequently, if the Big Bang took place everywhere in space, one wonders how temperature would have dissipated from $10^{11}$ degrees Centigrade down to 3 three billion degrees Centigrade in such a short period of time (i.e., 14 seconds).

Among other things, the law of entropy indicates that things move from an area of high temperature to areas of lower temperature. Yet, if the explosion took place everywhere in space, where were the areas of lower temperature into which the initial high temperatures flowed in such a short period of time?

The foregoing considerations do not necessarily demonstrate that Weinberg's description is inaccurate. Nonetheless, the manner in which the dissipation of temperature occurred remains unclear.

Weinberg indicates that following the Big Bang and during the period when the temperature of the universe dropped to three billion degrees, cosmological conditions became conducive to an increase in the rate of annihilations that took place among electrons and positrons relative to their rate of being created through photons and neutrinos. Moreover, he stipulates that this increased rate of annihilation over the rate of creation would have slowed, to some degree, the process of cooling down in the universe due to the amount of energy that would have been released during the process of particle-particle annihilation.

However, if the Big Bang resulted in particles fleeing away from one another everywhere in space, how does one know that particles would have been close enough to each other after 14 seconds of travel at accelerated speeds to be able to collide with, and annihilate, one another even if the temperature of the universe had dropped? To speak of a Big Bang that is taking place everywhere in space and, in the process, causing particles to achieve high -- possibly recessional -- velocities with respect to one another, entails a certain lack of clarity.

More specifically, the standard view of the universe at the present time is that everything is, supposedly receding from everything else, so if one extrapolates this back to the time of the Big Bang, how do particles that are receding from one another subsequently collide with and annihilate each other? On the other hand, if the meaning of the idea that space exploded everywhere during the Big Bang just means that particles scattered in every direction, why would high temperatures necessarily interfere with -- or lower temperatures facilitate -- the rate of collisions?

Since no one knows exactly what the Big Bang was, or what mechanism was involved, or how space exploded, or how that explosion unfolded in relation to space, or how long the Big Bang lasted, or how -- or if – the energy of the Big Bang was imparted to particles, therefore, one has difficulty developing a clear understanding of what might, or might not, have taken place within the first three minutes following the Big Bang. A great many assumptions are being made by Weinberg and others concerning the course of events during, and following, that alleged event.

Weinberg goes on to claim that inferences can be made concerning the density of the initial cosmic soup when the latter radiated at a temperature of $10^{11}$ degrees Centigrade. The figure he gives is: $4 \times 10^9$ times the density of water ... and this is more of an speculative estimate than it is a conclusion that is derived from empirical data and first principles.

Furthermore, he indicates that the proportion of, on the one hand, protons and neutrons to, on the other hand, electrons, positrons, photons and neutrinos, was in the order of: one neutron and one proton for every thousand million electrons, positrons, photons, or neutrinos. Weinberg claims the foregoing number reflects the character of the cosmic background radiation that was detected by Robert Wilson and Arno Penzias in 1964.

Later on in this book, the possible nature of cosmic background radiation will be critically explored in greater detail. For now, Weinberg's comments concerning density and the proportion of protons/neutrons to electrons, positrons, photons, or neutrinos will be permitted to stand with one proviso: one should keep in mind that there still is a cloud of unknowing surrounding the issue of just how abundant positrons might have been prior to the Big Bang.

At the end of the first three minutes following the initial Big Bang, the temperature of the universe supposedly had dropped to one billion degrees Centigrade. At that point, the temperature was sufficiently cool and the density of the cosmic soup was still sufficiently high (remember, particles have been flying away from one another, and, therefore, diluting the initial density of the cosmic soup) to permit protons and neutrons to begin to form complex nuclei such as heavy hydrogen (i.e., deuterium, one proton and one neutron), as well as helium (two protons and two neutrons).

At the three-minute mark, Weinberg believes that hydrogen nuclei are approximately three times more prevalent than helium nuclei. The rest of the universe consists largely of electrons, positrons, photons, and neutrinos ... although as noted earlier, whether, or not, positrons existed in any appreciable numbers at the beginning of the Big Bang or at the three-minute mark is an on-going question.

A few hundred thousand years after the Big Bang, Weinberg believes the universe would have cooled sufficiently for electrons to begin to be captured by the hydrogen and helium nuclei that had begun to assemble around the three-minute mark. Furthermore, through the presence of gravitational forces, the hydrogen and helium molecules would begin to be drawn together and coalesce into the cosmic seeds that eventually, would give rise to the formation of stars and galaxies.

Although millions of years would be required to get the universe moving in the direction of star and galaxy formation, nonetheless, according to Weinberg, everything that was necessary for those subsequent dynamics arose during the first three minutes following the initial Big Bang. However, while gravitation is believed to have begun playing a significant role in shaping the universe once hydrogen and helium molecules form and start to mutually draw one another together, nonetheless, the origins of gravity are still something of a mystery.

Einstein famously said that gravity was geometry, just as he said that time is what a clock measures. Nonetheless, he never explained how the geometry of space-time became warped in the first place, anymore than he explained how velocity was capable of affecting the ontology of time rather than just affecting the means of measuring time under certain conditions.

Gravity supposedly manifested its presence through the curvature of space-time, and the curvature of space-time marked the presence of gravity. Yet, although Einstein provided a way (i.e., General Relativity) for describing the dynamics that took place within a given gravitational context, Einstein couldn't explain how the space-time character of such a gravitational context initially came into existence.

General Relativity was all about describing the dynamics of existing gravitational fields. General relativity had nothing to say about how a gravitational field came to be in the first place, and this is probably the biggest reason why scientists have had such difficulty reconciling the force of gravity with the other three forces (electromagnetic, weak, and strong) because treating gravity as geometry doesn't permit one to examine the dynamics underlying the emergence of such geometry and, thereby, account for how the warping of space-time is possible (All of these issues will be explored in more detail later in the book.).

For Einstein, gravity is not a force per se. For him, gravity is a matter of how objects behave as a function of the way one field of space-time curvature affects, and is affected, by other fields of spacetime curvature and, consequently, the mutual "attraction" of two masses really gives expression to the relationship between their respective conditions of curvature ... a relationship that can be described through tensor calculations that keep tract of the manner in which different kinds of space-time curvature play off against one another.

In General Relativity, there is no mysterious action at a distance as there was with Newton. Different fields of curvature flow together in space-time, and gravitation gives expression to the way such flowing sets of curvature (i.e., gravitational fields) change character over time.

Steven Weinberg does admit that his outline of what happened during the first three minutes following the Big Bang has some lacunae. He also acknowledges there is a considerable lack of clarity present in our understanding about what might have been taking place during the first $1/100^{th}$ second of the Big Bang.

One might note in passing, that Weinberg (along with the rest of us) knows even less about what transpired prior to the on-set of that first $1/100^{th}$ of a second. And, due to such ignorance, we have no idea why the Big Bang occurred – if this is what actually took place – or what the Big Bang entailed ... if it did occur.

| Cosmological Frontiers |

The first three minutes is a narrative that weaves together certain strings of scientific understanding and, in the process, seeks to create an explanatory tapestry concerning the origins of the universe that we experience today. That narrative might turn out to be true – in part or in whole – but at the present time, Weinberg's account of the first three minutes is more of a story than it is necessarily an expression of some sort of unvarnished truth.

Part of Weinberg's story is rooted in scientific findings. However, part of that story is rooted in speculations concerning: What the scientific findings being alluded to actually mean, or how such findings might properly fit together, or whether those findings are even scientific in any indisputable sense.

Weinberg claims that the foregoing 'Standard Model of Cosmology' constitutes a consensus of scientists that has been forged through the shaping impact of empirical fact. However, as this book unfolds, the ensuing discussion will seek to raise many questions concerning the value of that consensus or the quality of the empirical facts through which that consensus supposedly has been established.

Pappus of Alexandria reports (in Greek) that Archimedes once uttered: "Give me a place on which to stand, and I will move the Earth" – an English translation of the original Greek. The statement was in reference to the potential power of the lever to move sizable masses if human beings utilize that instrument properly under various circumstances.

Of course, having a place on which to stand would not be sufficient. One also would need a place on which to place a fulcrum that constitutes a key element with respect to being able to pivot a lever in a manner that is capable of moving a mass, and as well, the composition of the lever would have to be able to withstand whatever stresses were placed on it while trying to move the Earth (not to mention the breathing gear and protective suit that would be needed to survive long enough to try moving the Earth from a vantage point somewhere beyond Earth.).

Whether, or not, Archimedes actually could have translated the limits of human strength in a manner that would have been capable of moving the Earth even if he had been given a place on which to stand and a properly placed fulcrum with which to work is an empirical question ... one that has never actually been put to the test. A similar sort of issue arises in conjunction with the Standard Model of Cosmology.

More specifically, scientists use assumptions to leverage theoretical movement. If one concedes that such assumptions are true, then through that concession, scientists acquire the capacity to manipulate almost any aspect of experience and move understanding in the desired theoretical direction.

In other words, granting such assumptions is comparable to providing scientists with a place on which to stand and a place on which to place their fulcrum, along with providing those scientists with the appropriate sort of lever and protective gear needed to stand where necessary for as long as necessary to accomplish such a process of leveraging. However, just as Archimedes' claim is difficult to put to the test because providing him with the conditions he requires to prove the truth of his claim is very difficult, if not impossible (at least at the present time), so too, it is exceedingly difficult (if not currently impossible) to provide the proponents of the Standard Model of Cosmology (i.e., the Big Bang) with the empirical and experimental conditions they need to prove their claim because such conditions are rooted in circumstances (i.e., the nature of the Universe prior to, and during, the Big Bang) that have become lost in the mists of, possibly, unknowable things.

One can start with the data that is present in today's universe and try to extrapolate back to how the universe operated billions of years ago. Unfortunately, there could be a multiplicity of theoretical avenues through which the current world might have assumed its present form, and this problem of multiple possibilities becomes especially difficult when one cannot establish the nature of the initial conditions from which everything supposedly arose ... something that Weinberg acknowledges to be the case – at least in certain respects -- as far as the first three-minute scenario is concerned.

Weinberg does point in the direction of a possible alternative to the Big Bang scenario when he briefly mentions several steady-state models that arose in the 1940s through the work of Thomas Gold and Herman Bondi, as well as through the efforts of Fred Hoyle. Weinberg's primary objection to such theories seems to be that those individuals discarded the problem of the early universe.

Nonetheless, the problem of the early universe only exists if, in fact, there was an early universe. Weinberg assumes that such a problem is real, whereas the proponents of steady-state theories assume that because the universe has always been present, there is no need to address the problem of an early universe ... for people like Gold, Bondi, and Hoyle, the latter matter is a non-issue.

Weinberg acknowledges the possibility that the Big Bang model ultimately might be proven to be either incorrect or replaced by some other set of ideas (e.g., a revised version of a steady-state model). For now, however, he believes that the Big Bang perspective is considered by many scientists to offer a standard frame of reference through which to filter various cosmological research proposals in an attempt to discuss, test and evaluate such proposals, and, moreover, Weinberg believes that if the day should come when the Big Bang theory gives way to a more heuristically valuable cosmological contender, he believes that the Big Bang model will have served a fulcrum-like role that the lever of continued scientific inquiry will have used to conceptually pivot the new theory into view.

The remainder of Steven Weinberg's *The First Three Minutes* runs through the foregoing outline in more technical detail as well as provides an array of considerations that are intended to lend scientific credence to the idea that the Big Bang model represents a good approximation of what began to unfold some 14 billion-plus years ago. Nevertheless, the questions and problems that have been raised during the last 10-15 pages of this chapter continue to persist despite Weinberg's attempt in the rest of *The First Three Minutes* to direct attention away from such weaknesses and toward what he considers the strengths of the Big Bang model.

The present book is not written from either a Big Bang or Steady-State perspective. I am more interested in exploring what those theories claim and, whether, or not, such claims are warranted. I also am interested in exploring the to extent to which those models can be said to constitute the sort of understanding that helps to adequately resolve the reality problem with which we are all confronted within the conditions of Final Jeopardy ... that is, having only the days of our lives (whether many or few) to come up with what we consider to be the best response that can be given to the reality problem that is contained with the Final Jeopardy challenge.

Chapter 2: The Meaning of Red

I brought a copy of Halton Arp's book: *Seeing Red: Redshifts, Cosmology, and Academic Science* in the late 1990s. Even when circumstances (and choices) required me to make a number of moves from: Canada, to: New York, New Jersey, and, finally, various places in Maine, I continued to hang on to that book, along with a variety of other volumes in which I was interested.

Eventually, life began to settle down. Nonetheless, for the next twelve years, or so, Arp's book served primarily as a surface for collecting dust.

When the idea of the *Final Jeopardy* series of books bubbled to the surface a few years ago (not too long after I finished writing *The Unfinished Revolution*), I wanted the aforementioned series to include a volume that would explore various aspects of cosmology, and, consequently, *Seeing Red* got dusted off and put in a queue of materials to be read.

Thirty years earlier I had read Steven Weinberg's *The First Three Minutes*. At that time, I made some notes concerning a few problems with the latter book, but I didn't begin to follow up on those sorts of conceptual issues until I started to outline how the *Final Jeopardy* series of books might unfold, and, as a result, some of the problems being alluded to above were finally given expression in the previous chapter of the present book.

Arp's aforementioned work examines a central theme of the Big Bang model ... namely, the idea that the greater the redshift frequencies are that are associated with the spectral properties of certain light sources in the universe (both stars and galaxies), then the more distant such light sources are relative to light sources that exhibit frequencies which are less shifted toward the red end of the spectrum. In addition, Arp also explored a related idea that objects displaying greater redshifts are receding from an observer in direct proportion to the magnitude of the measured redshift.

If the degree of spectral redshift is not an indication that: (1) objects are further away than objects that exhibit less of a spectral redshift, or that: (2) objects displaying greater redshift are receding away from us at velocities that are proportional to their redshifts (and, therefore, the greater the redshift, then the faster an object is receding away from us), then, the Big Bang model is confronted with a huge problem. The whole idea of the Big Bang revolves about the idea that the universe began to expand from the moment the primeval explosion took place that supposedly marked the beginning of our universe.

Since the time of Hubble, the existence of higher redshifts in the spectral properties of various stellar and galactic objects relative to the redshifts in the spectral properties of other stellar and galactic objects has been considered to constitute proof that the universe is expanding. Whether one talks in terms of a singularity that was ripped apart or a cosmic soup that exploded everywhere at once, the Big Bang is permeated with the properties of expansion.

Arp is willing to concede that, in some instances, redshift might be associated with the idea of distance. Nonetheless, he believes there are many instances in which redshift has nothing to do with either distance or recessional velocity.

He refers to such redshifts as being "intrinsic". In other words, these kinds of redshifts have something to do with the dynamic properties of the astronomical object being considered, and, consequently, do not arise as a function of increasing distance and/or recessional velocity.

What Arp believes such intrinsic redshifts entail will be touched upon later on in this chapter. For now, a little background is in order.

Hubble put forth his law concerning the relationship between dimness/faintness of astronomical objects (an indication of apparent magnitude) and redshift in 1929. Many astronomers – including, at times, Hubble himself -- interpreted his law to mean that the greater the redshift associated with an astronomical object, then the more distant (i.e., dim or faint) such an object should be considered to be.

More importantly, Hubble applied his law to the entire universe. Redshift became the key to understanding the rate and character of the phenomenon of universal, cosmic expansion. Everything was moving away from everything else.

As pointed out in the previous chapter, one of the first heavenly objects studied by Hubble was M31 (now known as the Andromeda galaxy). M31 exhibited a blue shift in its spectral properties, suggesting that the galaxy was moving toward us at a speed of approximately 100km/sec.

How could everything in the universe be expanding if objects such as M31 were moving toward us? Moreover, although redshifts seemed to predominate among the astronomical objects that were studied by Hubble, M31 was not the only object characterized by blueshifts in its spectral properties.

Furthermore, a number of scientists have pointed out that the correlation supposedly established by Hubble between redshift and distance/recessional velocities is suspect. For example, Steven Weinberg indicates in the *First Three Minutes* that almost no correlation can be established in Hubble's data between galactic velocities and distance, and, consequently, Weinberg proceeds to wonder if Hubble might have fudged the analysis of the redshift data to fit some preconceived idea Hubble had concerning the nature of the universe … for example, with respect to the issue of expansion.

In 1948, John Bolton discovered the existence of radio frequencies in the universe. Many of those radio sources tended to occur in pairs, and, as well, they seemed to be connected, like filaments, to a galaxy that was usually characterized by a lower frequency radio wave than the pair of radio sources that appeared to be located on either side of such galaxies.

Not too long after radio sources associated with star-like objects (which later were referred to as quasars … a contraction of "quasi stellar") were discovered in 1963, such sources began to be analyzed spectroscopically. What, initially, were considered to be odd spectra for stars came to be understood as emission lines from galaxies and the lines were exhibiting considerable shifts toward the red end of the spectrum.

The degree of redshift seemed to indicate that quasars were both very distant and appeared to be receding at velocities that were near to the speed of light. In addition, if such objects were really as distant as their redshifts seemed to indicate, then they had to be thousands of times more luminous than any previously observed extragalactic object.

While working at the Palomar observatory located in San Diego County, California, Halton Arp began to explore the foregoing objects to determine what, if any, connection those objects had to galaxy formation. During the course of his initial explorations (between 1961 and 1966), he generated a catalog consisting of 338 galaxies that he considered to be "peculiar".

The objects featured in his catalog constituted a sample of near-by galaxies that exhibited structural properties that seemed to be distortions, of one kind or another, involving elliptical and spiral galaxies. Arp believed his catalog provided observational data that could help astronomers understand how elliptical and spiral galaxies might have formed.

When Arp began to more closely study the actual astronomical objects that were depicted in his catalog, he discovered that pairs of radio sources tended to be associated with the peculiar galaxies that exhibited the most structural irregularities. Even more intriguing was the fact that some of the radio sources emanating from those galaxies turned out to be quasars.

What made the foregoing discovery intriguing had to do with the distance of such quasars. Contrary to the beliefs of many scientists, the quasars studied by Arp seemed to be relatively close by (astronomically speaking).

Astronomers had been in the habit of interpreting the large redshifts associated with quasars to be an indication of their tremendous distance from Earth, as well as a reflection of their extremely high recessional velocities. Yet, Arp was claiming that some quasars with high redshifts actually existed relatively close to the Milky Way, and, therefore, the presence of such redshifts was not necessarily an indication of either tremendous distances or high recessional velocity.

Somewhat ironically, the person who had given Halton Arp his first job at the observatory in Pasadena was none other than Edwin Hubble, the godfather of redshifts. Arp's assigned task was to help establish the distance scale being used in cosmology.

Arp notes that in Hubble's book, *The Realm of the Nebulae*, Hubble said: "...if the interpretation as velocity shifts is abandoned, we find in the redshifts a hitherto unrecognized principle whose implications are unknown." Arp's discovery of quasars with high redshift values that were relatively close to the Milky Way galaxy were empirical instantiations of Hubble's speculative puzzlement about what redshifts might mean if they were not indications of velocity shifts in astronomically distant objects.

In his book, *Seeing Red*, Arp talks about a paper that was submitted by Wolfgang Pietsch in the hope that it would be published in *Astrophysical Journal Letters*. Pietsch's paper was rejected because, apparently, it gave expression to observational data that presented problems for the idea that redshifts were an indication of increased distance and recessional velocities.

More specifically, Pietsch's article was about the discovery of two X-ray sources that were paired across a galaxy known as NGC4258 (cf., The New General Catalog of Nebulae and Clusters of Stars). NGC4258 is a highly active Seyfert spiral galaxy (a class of galaxies first discovered by Karl Seyfert, an American astronomer, in the 1950s).

Pietsch claimed that the X-ray sources were quasars. They displayed a redshift that was far greater than the galaxy – i.e., NGC4258 – with which the X-ray objects were paired.

The referee who reviewed Pietsch's paper indicated that spectral analysis needed to be done with respect to the X-ray sources in order to confirm that they were, indeed, quasars. Moreover, the referee also wanted the redshift values for the X-ray objects to be verified.

In order to provide the additional data, Pietsch requested some time on the telescope at a certain European observatory. The request was denied.

Several years later, astronomer, Margaret Burbidge (she had been a past president of the American Association for the Advancement of Science as well as the Director of the Royal Greenwich Observatory) obtained the requisite information concerning the spectra of the X-ray sources first discovered by Wolfgang Pietsch. Her spectral observations demonstrated the similarity between the two quasars in Pietsch's original study.

She submitted her findings to the *Astrophysical Journal Letters*. A referee took exception to her observations, and, consequently, her study did not fare any better than the paper by Wolfgang Pietsch.

Most astronomers were inclined to interpret the findings of Pietsch and Burbidge as being nothing more than isolated cases that involved accidental alignments of astronomical objects. Such chance-alignments gave the illusion of connection when, according to such astronomers, there was none.

Did the foregoing astronomers demonstrate that the observations of Pietsch and Burbidge were accidental alignments that were illusory in nature? No, they didn't.

Astronomers merely assumed this was the case. Apparently, they believed their assumptions should be able to trump actual observations and empirical analysis.

Arp wrote a paper that encompassed a variety of calculations concerning how likely it was that the observations of Pietsch and Burbidge were of a chance, accidental, illusory nature with respect to the possible connection between galaxy NGC4258 and the X-ray sources/quasars that appeared to be paired with that central galaxy. Arp concluded that the likelihood that the foregoing observations merely constituted a chance alignment of astronomical objects was less than one in 2.5 million.

Arp's foregoing paper was not rejected in any outright fashion. It was merely shelved indefinitely.

At no point was evidence put forth indicating that Pietsch was wrong with respect to his original claims about the X-ray sources associated with NGC4258 being quasars whose redshift was greater than the galaxy with which they were paired. At no point was evidence put forth indicating that Margaret Burbidge's spectral analysis of those objects was incorrect. And, finally, at no point were Arp's calculations shown to be error-ridden.

Instead, gatekeepers operating out of the institutional establishment of astronomy stonewalled the observations and data of all three individuals. Such stonewalling was not rooted in what those gatekeepers could empirically demonstrate concerning the supposed flaws inherent in the work of the aforementioned three individuals, but, instead, those gatekeepers were luxuriating in the power that anonymous ideologues can wield with respect to what will, and will not, see the light of day in so-called scientific publications.

Arp, Pietsch, and Burbidge might, or might not, be correct with respect to their respective claims concerning the X-ray sources associated with NGC4258. However, those claims need to be disproven empirically and not just arbitrarily rejected because someone doesn't like the implications such claims carry with respect to some beloved pet theory about the meaning and significance of redshifts.

Were Pietsch's observations involving X-ray sources in conjunction with NGC4258 illusory in nature? If astronomers consider such cases as being isolated instances of chance alignments, then, perhaps the best way to build a case tying paired x-ray sources/quasars with a central galaxy is to find other instances of a similar nature, thereby, indicating that the original case was not an isolated case, and, in addition, the more instances one can cite that are of a similar nature, then, the more unlikely it becomes that all such examples merely involve chance alignments of astronomical objects.

Consequently, Arp began to compile an array of examples showing that highly X-ray sources/quasars with high redshifts were often paired with a central galaxy that exhibited a lower redshift. However, one of the primary points of contention concerning his examples was whether, or not, the indicated X-ray sources actually exhibited a physical connection to a central galaxy.

According to Arp, one could point to radio lobes or filaments that bridged the space between X-ray sources or quasars with a central galaxy. Other astronomers denied that such bridges existed.

For example, consider the quasar known as Markarian 205 (Mark205 ... related to the work of American astronomer, B.E. Markarian) that has been linked to an extremely active galaxy identified as NGC4319. The tentative connection between the two was first noted in 1971, but that linkage has been the subject of some controversy for more than a quarter century.

In 1990, the Max-Planck Institute for Extraterrestrial Physics launched an X-ray telescope known as ROSAT (Röntgen Observatory Satellite Astronomical Telescope). Arp submitted a proposal to the project that was geared toward uncovering evidence that Mark205 and NGC4319 were physically connected by radio sources.

Although Arp's time on ROSAT did not provide the evidence he was looking for, he did find something that was consistent with his underlying theory that quasars are paired with, and connected to, a central galaxy. More specifically, he discovered there were X-ray filaments emerging from both sides of Mark205 that terminated on X-ray sources that were later determined to be quasars.

Subsequently, Arp found spectral data relevant to the foregoing discovery. The data indicated that two quasars exhibiting redshift values of .63 and .45 respectively were connected to an object with a low redshift of .007.

When presented with the foregoing observations and data, some astronomers dismissed Arp's findings as just being a function of noise that Arp was misinterpreting as some sort of connection between two quasars. Yet, once again, those sorts of negative responses to Arp's material were not backed up with empirical data or analysis ... just a categorical rejection of a speculative nature.

In 1994 Arp had an unusual opportunity to participate in a four-day symposium of the International Astronomical Union. His participation in such a venue was unusual because ever since he had released his *Atlas of Peculiar Galaxies* in 1966, he had become an outcast and was denied access to the telescopes at all of the major observatories in America, eventually forcing Arp to leave America and continue his explorations at the Max Planck Institute for Astrophysics in Munich, Germany.

Arp used his opportunity at the 1994 IAU meeting to briefly outline some of his findings concerning the way in which high redshift quasars were paired with and physically linked with low redshift central galaxies. His perspective did not go unnoticed.

Toward the end of the symposium, and as was usual during such gatherings, one, or another, expert on astronomy was given the honor of presenting a state of the universe address that supposedly summed up what was currently known about astronomy. The presenter on that occasion was Martin Rees, an internationally renowned astrophysicist and cosmologist.

During his talk, Rees pointedly criticized Arp's relatively short presentation that had taken place a few days earlier. Following Rees' address, Arp was afforded a chance to make some additional remarks vis-à-vis his own position, and Arp used that opportunity to introduce some additional data and observations that he had not been able to include in his earlier talk ... observations and data that ran contrary to the standard model of cosmology as far as issues of redshifts and expansion were concerned.

Following the foregoing presentations, the meeting was opened up to questions. One journalist directed a question to Martin Rees and wanted to know why no one in astronomy seemed to be following up on Arp's observations and data.

Rees is reported to have responded to the question by making a vitriolic, personal attack on Arp. That is, instead of citing facts and putting forth cogent arguments that would demonstrate the errors in Arp's perspective, Rees mounted what seemed to be a baseless attack on Arp's competency as a scientist.

I've seen the foregoing tactic used in discussions involving: HIV, Antineoplastons, evolution, neurobiology, quantum physics, and pharmacological approaches to mental disorders (and, indeed, I have touched upon this issue in both *Volume I* and *Volume II* of *Final Jeopardy*). People who have vested interests (financial, professional and/or ideological) to protect often seem more interested in attacking – if not trying to shame -- those individuals who will not submit to the world-view of the former "experts" than they are interested in engaging in a mutually respectful search for the sort of truths that might be in everyone's interests.

As previously noted, one of the criticisms various astronomers directed toward Arp's findings was that his observations were merely a matter of chance alignments of astronomical objects in which high redshift, distant quasars were being erroneously linked with low redshift galaxies that were much, much closer to the Milky Way. The source of the error was allegedly due to the way Earth was situated relative to the extragalactic high-redshift quasars and the low-redshift galaxies being considered, and if one were able to view those quasars and galaxies from a very different angle independent of Earth, then, the connection being asserted by Arp would vanish and shown to be the illusion that such astronomers claimed them to be.

Unlike Arp's perspective, the foregoing sort of criticism was not backed up with any evidence. Instead, it was rooted in a hypothetical ... namely, if one changed the angle of observation, then, Arp's claim would be disproven, but this hypothetical was never tested.

In 2004, Margaret Burbidge presented a paper at a meeting of the American Astronomical Society that was being held in Texas. The paper was a collaborative effort by Burbidge, her husband, Geoffrey, Halton Arp, and several other astronomers.

The paper discussed the relationship between a high redshift quasar and a low redshift galaxy, NGC7319. This time no one could raise hypothetical possibilities concerning illusory alignments involving a high redshift quasar and a low redshift galaxy caused by Earth's relative position of observation with respect to such objects because the high redshift quasar was in front of the low redshift galaxy.

If redshift is an indication of distance and recessional velocity, then, the evidence being reported by Burbidge and others in the 2004 paper constituted a major problem for the Standard Model of Cosmology. For instance, why does a highly redshifted quasar appear to be closer to Earth than a relatively low redshifted galaxy, and how does one reconcile such data with a supposedly expanding universe whose expansion is tied to measurements involving redshifts that – in the light of the foregoing paper -- do not necessarily serve as reliable indicators of either distance or recessional velocity?

When the foregoing paper was submitted for publication, the individuals who were reviewing the paper made a strange demand. They wanted the authors to include a section in the article that offered a different account of the data being presented in the paper – namely, that the quasar was actually behind galaxy NGC7319 ... something for which there was no observational evidence but that did help to save appearances as far as the Standard Model of Cosmology is concerned.

In 2005 Dr. Martin Lopez-Corredoira publically summarized research that he had been conducting over a period of five years. The occasion was the First Crisis in Cosmology Conference.

His research was focused on the way in which galaxies with disparate redshifts were often associated with one another. If true, Dr. Lopez-Corredoira's findings would be a further indication that redshift was not necessarily a reliable marker of distance, recessional velocity, or an expanding universe.

His findings were consistent with the previously discussed work of Halton Arp and Margaret Burbidge. Among other things, he discovered there were empirical indications that some quasars and galaxies were closer to the Milky Way galaxy than their redshift values would suggest, and, as well, he found that there were radio and X-ray filaments that seemed to link astronomical objects exhibiting different redshifts.

Dr. Lopez-Corredoira experiences during his five-year period of research also mirrored the work of Halton Arp in, yet, another way. After being granted a few nights of observation time between 2001 and 2002, Dr. Lopez-Corredoira was ostracized from observatories just as Halton Arp had been denied access following the publication of his *Atlas of Peculiar Galaxies* in 1966 ... apparently, the astronomers who controlled the use of telescopes didn't want anyone using their equipment who might generate observations and data that departed from the worldview of the Standard Model of Cosmology.

Galileo had invited Church officials to look through his telescope to be able to see the nature of things for themselves. Apparently, the high priests of the Church of Standard Model Cosmology wanted to discourage people from looking through their telescopes unless they did so through the filters of a set of preconceived notions.

There is another dimension of redshifts that is both intriguing and mysterious. Dr. Margaret Burbidge and her husband, Geoffrey, had discovered in 1967 that there was a certain value of redshift – i.e., $z = 1.95$ – that appeared to be preferred by many quasars.

The foregoing finding induced K.G. Karlsson to further pursue the issue. In 1971, he derived a formula – $(1 + z_2)/(1+ z_1)$ – in conjunction with his studies of quasar redshifts that appeared to indicate that those values were quantized and generated a series consisting of the following values: 0.061, 0.30, 0.60, 0.91, 1.41, 1.96 ... .

The last entry in the above series is very close to the value that Margaret Burbidge and her husband had noted in their 1967 study of redshifts. Is this a coincidence or is it an indication of the existence of some sort of deep principle that governs the values of redshifts among different quasars?

A quarter century later, in 2006, M.B. Bell and D. McDiarmid, working at the National Research Council of Canada, released a study that involved an analysis of the redshift values of more than 46,000 quasars that were listed in the Sloan Digital Sky Survey. The two researchers reported there were peaks in the distribution of redshifts among the quasars they studied that were consistent with the series of redshift values that were given through the intrinsic redshift equation noted earlier.

No one is quite sure what any of the foregoing findings ultimately mean. However, what such data does appear to indicate is that quasar redshift values are, for some reason, quantized as a function of something that is intrinsic to those quasars and that is not a function of either distance or recessional velocity.

According to Halton Arp, the intrinsic redshift values displayed by quasars are an indication that galaxies do not necessarily form as a result of the gravitational aggregation of gasses and cosmic dust that supposedly takes place over millions of years. Instead he believes that older, more mature galaxies eject massive, highly luminous quasars in pairs that serve as the seeds for the emergence of future galaxies.

The youthful quasars exhibit a greater redshift relative to the older galaxies from which they arise. Over time, the redshift values associated with such quasars tend to decline.

Moreover, the quasars that are ejected continue to have a physical relationship of some kind with the parent galaxy. For Arp, the proof of such a continuing relationship is found in the radio and X-ray filaments that he has detected which appear to bridge the space between the quasars and the parent galaxy.

Finally, the foregoing quasar-dynamics are most likely to occur in relation to galaxies that display the greatest amount of physical or energetic disturbance. These are the sorts of galaxies that tend to populate the pages of Arp's *Catalog of Peculiar Galaxies* and that were likely to contain the Active Galactic Nuclei through which quasars were, somehow, formed and from which the latter objects were, somehow, ejected.

There are two separate issues concerning redshifts that are inherent in the work of Halton Arp. One issue has to do with whether, or not, redshifts represent the birth of new galaxies via ejected quasars that continue to maintain a physical relationship with the parent galaxy via radio and X-ray filaments.

The other issue is whether, or not, redshifts constitute a reliable indicator of distance and recessional velocity. This latter issue can be considered quite independently of the matter of galactic formation ... that is, one could reject Arp's theory about galactic formation and still accept his observations indicating that astronomical objects of disparate redshift values often associate together, thereby demonstrating that redshift values are not necessarily a function of either distance or recessional values.

By his own admission, Arp does not know how parent galaxies are able to form or eject massive quasars, nor does he know why this happens. In addition, although elevated redshifts are associated with such "ejected" objects, what that redshift signifies is not known.

In addition -- and for the sake of argument -- one could accept Arp's claim that there are X-ray and radio filaments physically linking a central galaxy to the quasars that are paired with it. Nonetheless, conceding such a point does nothing to explain what is transpiring along, or through, those filaments.

Finally, one might be willing to accept the studies that have been conducted by the Burbidges (husband and wife), K.G. Karlsson, as well as the Canadian researchers Bell and McDiarmid – all of which indicate that intrinsic redshifts seem to gravitate toward certain preferred, quantized value. Nonetheless, no one knows what this means.

If one knew what intrinsic redshift meant, one might have some insight into why its values seem to be quantized. However, absent that sort of understanding, those redshift values just give expression to correlations of unknown significance.

None of the foregoing considerations are intended to suggest that Arp's theory about galactic formation is incorrect. Rather, the preceding discussion merely raises some points indicating that, currently, Arp's theory entails some important lacunae that, eventually, might, or might not, be filled in with the sort of data and observations that are capable of substantiating his basic premise concerning the possible role that quasars, filaments, and redshifts might play in the formation of galaxies.

The data being alluded to in the foregoing paragraph is to be found – or not – in the future. However, one does not need to look to the future to appreciate how Arp's work (along with the efforts of Wolfgang Pietsch, the Burbidges, Dr. Lopez-Corredoira and others) already strongly suggests that redshift is not necessarily an indication of distance, recessional velocity, and/or expansion.

To say, on the one hand, that redshifts do not necessarily signify distance does not mean some part of a redshift's value couldn't be due to the effect that distance has on the frequency of light emanating from astronomical objects. On the other hand, however, the idea that redshifts aren't necessarily tied to distance (for example, in the case of intrinsic redshift) means that one has to disentangle the contributions of distance-related factors from non-distance related factors when making various kinds of cosmological calculations.

Moreover, even when one is able to determine what the distance-related contribution might be with respect to some given redshift, this does not necessarily mean that such distance is a function of, or can be equated with, recessional velocity and, therefore, an expanding universe. For example, consider the issue that is referred to as "tired light".

The space between galaxies might be referred to as voids, but they are not empty. Various forms of energy (e.g., zero-point energy) manifest themselves in the vacuum of space, and, as well, such spaces contain cosmic dust and atoms of gas ... especially hydrogen.

Although clouds of hydrogen have been observed to be fairly ubiquitous in intergalactic space, there might be much more hydrogen in those regions than we know. This is because molecular hydrogen does not give off radiation that can be detected.

When photons pass through the foregoing sorts of relative voids, they might interact with atoms and molecules in those regions. When this occurs, the photons often: Are absorbed by such atoms or molecules, lose a certain amount of energy in the process, and, then, are emitted again to continue on their way without any change in direction.

How do we know that the foregoing sequence of events might occur? The 1988 work of the Canadian scientist, Paul Marmet (in conjunction with the American-born radio astronomer, Grote Reber) and, independently, the investigations of Jacques Moret-Bailly from France (the CREIL effect), tend to indicate as much.

The foregoing researchers used different methods. Nonetheless, they each discovered that there seemed to be a non-Doppler shift toward the red end of the spectrum that took place when light interacted with atoms and molecules. The presence of such a redshift (after the contribution from other factors had been removed from the calculations) reflected the loss of energy undergone by photons during their interaction with various atoms or molecules.

As light travels across the universe from some cosmic source toward the Earth, it becomes older. The older light is, the more distance it has covered in its journey to Earth.

The older light is – that is, the greater the distance it has traveled – the more space it has traversed. The more space that has been traversed, then, the more likely it is that such light will have encountered atoms, molecules, dust, and various energy fields.

With each encounter, the spectral properties of that light will be shifted toward the red end of the spectrum as a result of the energy that is lost through the foregoing kinds of interactions. Such redshifts are non-Doppler in character ... that is, they are not caused by recessional velocities, but, instead, they are due to the loss of energy during interactions with atoms, molecules, and so on, along the journey toward Earth.

Noble-laureate Richard Feynman rejected the idea that the redshift that is found in the spectral properties of light that reaches Earth from distant, cosmic sources is due to "tired light". He argued that light interacting with atoms and molecules in intergalactic space would be scattered via the Compton effect, and, as a result, distant images would be blurred (which is not the case).

The research of Marmet, Reber, and Moret-Bailly indicates that Feynman might not be right. In other words, there is evidence to suggest that light might just lose energy, rather than be scattered, when that light encounters atoms and molecules in intergalactic space.

Conceivably, light might be scattered under some circumstances, while under other conditions, light might be absorbed, lose energy, and, then, be re-emitted with a slight increase in redshift in its spectral properties. What such differential circumstances might involve is unknown, and what is also unknown is what actually happens to light as it travels across space on its ways to Earth because no one is traveling along with such light to determine what occurs at each point in the journey.

The perspectives of Feynman as well as Marmet and Moret-Bailly are each rooted in a certain amount of empirical data. However, each of those perspectives is a function of interpretations concerning what different aspects of the available evidence seem to suggest, and, therefore, which, if either, side of the "tired light" issue is correct is uncertain.

However, the "tired light" issue can be considered independently of the data involving quasars and galaxies that seem to be physically connected in some way and, yet, display disparate redshift values. More specifically, even if turns out that light does not lose energy as it travels across the universe and, therefore, this will not be reflected in its redshift values, nonetheless, those who are advocates of an expanding universe have to be able to explain why a quasar with a high redshift value has been located in front of a galaxy with a low redshift value if redshift is an indication of distance and recessional velocities.

The anonymous reviewers of the aforementioned paper by Margaret Burbidge, Geoffrey Burbidge, Halton Arp and others indicated that the authors needed to include a reference to an alternative account for the observation being discussed in their paper – namely, that the high redshift quasar in question was actually behind the low redshift quasar. Such a request seems ludicrous unless those reviewers can provide a plausible account of why what the authors were claiming should be considered to be illusory and that what had been observed with respect to the high redshift quasar and low redshift galaxy was not the case.

Of course, the reviewers being alluded to in the foregoing paragraph might have in mind something like the Einstein Cross, a system of five objects that appear to be related. This system – also known as G2237+0305 – is considered one of the best examples of gravitational lensing that has been discovered so far.

Gravitational lensing involves the idea that some massive cosmic object such as a galaxy or cluster of galaxies could be aligned with a more distant cosmic object, such as a quasar, in a way that caused the background object – e.g., quasar – to appear as if it were an image in the foreground or, alternatively, generate multiple images of the background quasar in the form of what is known as an Einstein Ring of objects that appear to encircle the source of the gravitational lensing (e.g., a galaxy or cluster of galaxies). However, if the components of an Einstein Ring are, indeed, images that have been created by just the right sort of alignment involving: (1) A background object (e.g., quasar), (2) a foreground source that gives rise to gravitational lensing (e.g., massive galaxy or cluster of galaxies), and (3) a point of observation that is properly aligned with the foregoing two elements (e.g., an observatory on Earth), then all of the images that are created should have the same optical properties, but this is not always the case.

Moreover, quasars exhibit a high proper motion. Proper motion involves the rate of angular change in an object's position over time as viewed from a given frame of reference – say, an observatory on Earth.

Consequently, despite the vast distances involved, a given alignment between a background object and a foreground gravitational lens cannot be maintained for longer than a few decades relative to an observatory on Earth. Yet, more than four decades of recorded data that have been compiled in conjunction with quasars indicate that such alignments have continued to retain their relationships and, therefore, this seems to suggest that what is being observed in relation to those alignments is not a function of gravitational lensing.

In addition, Arp published research in 1998 indicating that the connection between the four quasars in the Einstein Cross and the central galaxy was of a material rather than something generated through the process of gravitational lensing. Arp's evidence for making such a claim was rooted in spectroscopic analysis of the Einstein Cross system ... a claim that was independently and spectroscopically corroborated by Howard Yee.

The current chapter has explored, rather briefly, only a very small sampling of the research that has been conducted by Halton Arp across more than 50 years of work. The material selected here was intended to serve as a means of giving expression to some of the basic themes of Arp's perspective involving objects with disparate redshifts that appear to be physically connected to one another and the possible implications that such putative relationships carry for the idea of an expanding universe and accretion theories concerning the formation of galaxies.

At a minimum, Arp's work brings into question the meaning and significance of redshift values. If cosmic objects of disparate redshifts are physically linked with one another, and/or if quasars with high redshift values are calculated and observed to be closer to the Milky Way than are galaxies with much lower redshift values, then, there is a very strong possibility that redshift cannot be considered as a reliable indicator of either distance or recessional velocity, and, if this is the case, then a major piece of alleged evidence has been removed from the perspective that claims we live in a universe that is expanding.

The foregoing statement does not prove that the universe is static. Rather, it merely claims that if the universe is expanding, then, one cannot necessarily rely on redshift values as proof of that expansion.

Chapter 3: Noise

In 1964, Robert Wilson and Arno Penzias were engaged in the science of radio astronomy. They were working with a Bell Telephone Laboratory radio antenna located in Crawford Hill in Holmdel, New Jersey.

The antenna involved a 20-foot horn reflector that generated very low noise. Although the antenna originally had been constructed to operate in conjunction with Project Echo (source of the first two communications satellites ... 1960 and 1964), Wilson and Penzias were working with the antenna to try to measure the amount of radio waves emanating from the Milky Way galaxy, and the antenna's quality of low noise would be an asset in their investigations.

The foregoing sorts of measurements are not easy to make. To do so, one has to be able to differentiate between, on the one hand, radio waves from the Milky Way and, on the other hand, the electrical noise that is produced through the random movement of electrons in relation to the antenna, itself, (for example, through its amplifier circuits), along with whatever radio noise in the atmosphere that arises from Earth-based activity that might be picked up by the antenna.

The wavelengths in which Wilson and Penzias were interested were located at 7.35 centimeters and 21 centimeters. While the foregoing lengths give expression to radio waves, they also are known as microwave radiation.

The first wavelength – 7.35 centimeters -- was selected in order to determine how much electrical noise might be generated through the antenna apparatus itself. At 7.35 centimeters, the amount of radio noise that would be coming from the Milky Way should have been virtually non-existent, and, consequently, whatever radio noise was being picked up at this wavelength was likely due to electrical noise coming from the antenna, and, once known, could be factored in when making their observations and calculations.

The two radio astronomers were not expecting much radio noise to be generated by the antenna. Nonetheless, they had to make sure this was the case and not merely assume it to be so.

| Cosmological Frontiers |

Once they had calibrated things at the 7.35-centimeter wavelength, they planned to move on to looking at radio noise sources at 21 centimeters. The latter wavelength was considered to be prime hunting grounds for finding radio noise coming from the Milky Way galaxy, and, having studied things already at 7.35 centimeters, they would be in a position to be able to distinguish between radio noises produced by, on the one hand, the antenna and/or Earth-related radio noises and, on the other hand, radio noises coming from the Milky Way.

Unexpectedly, Wilson and Penzias discovered a significant level of radio noise at 7.35 centimeters. The noise seemed to be unrelated to direction, time of day, or season.

Lack of correlation with direction was especially important. If the radio noise were directional in character, it likely would have arisen in conjunction with some radio noise on Earth, but this did not seem to be the case.

After running through various possibilities and rejecting them as causes of the radio noise at 7.35 centimeters, the two radio astronomers re-visited the issue of how much radio noise might be generated by the antenna. As a result, the antenna was partially disassembled and cleaned.

None of the foregoing steps produced appreciable changes in the level of radio noise being received. The source of the unexpected level of radio noise at 7.35 centimeters continued to elude Wilson and Penzias.

The radio noise that was being detected at 7.35 centimeters was translated into a temperature by the two investigators. This is a standard practice among radio engineers.

A container with opaque sides and that exists at any temperature above absolute zero will generate radio noise as a function of the thermal motion of the electrons within such a container. The higher the temperature of the container's walls, the more radio noise will be emitted.

Thus, every level of radio noise is associated with a temperature that would be necessary to generate such an amount of radio noise in a container with opaque sides. The temperatures associated with the level of radio noise being received by the Bell antenna were determined by Wilson and Penzias to be between 2.5 and 4.5 degrees above absolute zero (Kelvin).

Relative to what they had been anticipating prior to gathering data through the Bell radio antenna, both Wilson and Penzias were puzzled about why the temperature associated with the 7.35 centimeter radio noise was as high at it was. In addition, they also continued to puzzle over what the source of the radio noise might be that they were observing.

Due to their puzzlement, they put off publishing their data right away. Eventually, however, despite not being able to resolve their puzzlement concerning the radio noise, they published their results.

At some point following the release of the foregoing data, Penzias made contact with a fellow radio astronomer, Bernard Burke. Burke told Penzias about a friend of his who recently had heard a lecture given by P.J.E. Peebles, an astrophysicist from Princeton University.

During the course of Peebles' talk, the speaker indicated there should be a remnant of radio noise left over from the Big Bang. The temperature equivalent cited by Peebles was calculated to be in the vicinity of 10 degrees Kelvin.

According to Peebles, the existence of radiation following the Big Bang played a crucial role in the unfolding of the universe. More specifically, if there had not been a very high level of radiation present during the first several minutes following the Big Bang, nuclear reactions would have taken place so quickly that a very large proportion of the hydrogen that existed in the early universe would have been transformed into heavier elements.

Since hydrogen seems, currently, to make up roughly 75% of the elements present in the universe, then, apparently, the process of nucleosynthesis could not have taken place to any appreciable degree during the period immediately following the Big Bang. One way to account for why such a process did not proceed in rapid fashion following the Big Bang is that the presence of high levels of very energetic, short wavelength radiation prevented that process of nucleosynthesis from taking place by keeping nuclear elements separated from one another.

There is, however, another possible explanation for why hydrogen currently makes up such a large proportion of observable matter in the universe. This alternative explanation does not require the presence of a large amount of radiation to prevent nucleosynthesis from taking place in relation to hydrogen, but, instead, is rooted in the idea that the Big Bang never took place, and, therefore, at no time did the necessary conditions of temperature, and so on, exist in the universe that would have resulted – were it not for the presence of large amounts of radiation -- in the rapid transformation of a large proportion of available hydrogen into heavier elements independently of the process of nucleosynthesis that can take place both within, as well as outside of, stars.

As the universe expanded following the Big Bang, the temperature of the universe supposedly fell. The fall in temperature would have been in inverse proportion to the size of the universe.

According to Peebles, the universe today (meaning the time when Peebles gave his lecture) continues to show telltale signs of the Big Bang and the ensuing dynamic between radiation and matter occurring at that time. One of those signs is in the form of the temperature equivalent of the radio noise that emanated from the events of long ago ... and, according to Peebles, the aforementioned 10 degrees Kelvin is the temperature equivalent of the radio noise that forms the background radiation that links the present universe to the time of the Big Bang.

Prior to Peebles, George Gamow had made a calculation concerning the same sort of background radiation. The figure Gamow came up with was 50 degrees Kelvin, five times higher than Peebles figure.

In 1947, two of Gamow's collaborators – Ralph Alpher and Robert Herman – made some calculations of their own with respect to the background radiation issue and came up with a figure of 5 degrees Kelvin for the temperature that was the equivalent for the background radiation that still pervades the universe and, supposedly, is a remnant of the Big Bang. A few years later, Alpher and Herman reworked their previous calculations and came up with the figure of 28 degrees Kelvin.

More than 20 years earlier, Arthur Eddington had made some calculations concerning the temperature equivalent of the background radiation that might exist in the observable universe. His calculations were not rooted in the dynamics of a Big Bang but, instead, they were based on trying to estimate the ambient, cosmic temperature of the universe when contributions from all the galaxies and matter of the known universe were taken into account.

The result of the foregoing calculations was 3.2 degrees Kelvin. This result is closer to the observations of Wilson and Penzias than it is to the calculations of Gamow, Peebles, and, with one exception (the aforementioned 5 degrees Kelvin), the calculations of Alpher and Herman, and, yet, Eddington was approaching things from a perspective that treated background radiation as a function of something other than what might be left over from a Big Bang.

In the 1930s, one individual (Ernst Regener) calculated that interstellar space would end up with a certain temperature at equilibrium. His perspective is somewhat similar to that of Eddington, but it is stated in a slightly different way.

Regener's equilibrium figure was 2.8 degrees Kelvin ... just .4 degrees Kelvin away from Eddington's aforementioned calculation. Regener's result is also closer to the observed data of Wilson and Penzias than it is to any of the calculations – with one exception (i.e., the 5 degrees Kelvin calculation of Alpher and Herman) – that were made by individuals who were advocates of a Big Bang theory.

In 1941, a Canadian physicist, Andrew McKellar, made some observations and related calculations with respect to background radiation and did so independently of any Big Bang considerations. The figure he arrived at was 2.3 degrees Kelvin, and this was comparable to the results achieved by both Eddington and Regener prior to him.

Irrespective of the issue of whether any of the foregoing calculations are correct or accurate, an important consideration needs to be kept in mind. The temperature equivalency spoken of by Wilson and Penzias might well give expression to a remnant left over from the Big Bang, but, as well, their data might reflect the radio noise generated by the movement of electrons in interstellar space as a function of the collective interaction of all matter/energy within a given volume of space and the impact that sort of dynamic has on the ambient, equilibrium temperature of that region of space.

Initially, Wilson and Penzias were unable to form a plausible hypothesis concerning the significance or meaning of the observations they had made in conjunction with the radio antenna at Crawford Hill. Things, however, were about to change.

During the aforementioned conversation between Penzias and his fellow radio astronomer, Bernard Burke (a conversation that dealt with, in part, information concerning the talk by Peebles and the latter's estimate concerning the magnitude of the temperature equivalent that should be detectable today and which was a cooled-down remnant of the radio noise that would have existed at the time of the Big Bang), Burke made a suggestion to Penzias. Perhaps Wilson and Penzias should contact Peebles and see what he might have to say about their data.

Penzias followed up on Burke's suggestion and phoned Princeton. However, he reached Robert Dicke instead of Peebles.

Dicke was a senior experimental physicist working at Princeton, and he had created some of the fundamental techniques used by radio astronomers in conjunction with the study of microwave radiation. Dicke's work actually played a role in helping to lead to Peebles' calculations concerning cosmic background radiation.

While Wilson and Penzias were busy with the study of radio noise in one part of New Jersey, Dicke was busy in another part of New Jersey meditating on the possibility that there might be radiation connected with an early, hot universe that still could be detected in 1964. He thought the issue was worth investigating and induced two other physicists – D.T. Wilkinson and P.G. Roll – to begin looking into the matter.

Before the foregoing research had been completed, Penzias – as noted earlier -- made contact with Dicke. On the basis of that conversation, a decision was made to submit two letters in tandem to the *Astrophysical Journal*.

Wilson and Penzias would author one of the letters. Dicke, Wilkinson, and Roll would write the other letter.

The letter by Wilson and Penzias outlined their basic research and data. However, their letter also contained a fleeting reference to the idea that one possible explanation for their findings was the perspective being outlined by Dicke, Wilkinson, and Roll in the latter's letter to the same *Astrophysical Journal*.

When Wilson and Penzias wrote their letter, they still were not completely sold on the idea that their observations meant what Dicke, Wilkinson and Roll were indicating in their own letter to the Astrophysical Journal. Nonetheless, the two researchers were prepared to acknowledge that the perspective of Dicke and his colleagues might account for the data gathered by Wilson and Penzias.

Proponents of the Big Bang theory claim that the equivalency temperature – roughly 3 degrees Kelvin -- determined by Wilson and Penzias in relation to the radio noise found at 7.35 centimeters is just what one would expect if the universe expanded by a factor of 1000 from the point in time when the temperature of the universe was sufficiently high to maintain thermal equilibrium between radiation and matter. The necessary temperature for the foregoing sort of thermal equilibrium is 3000 degrees Kelvin (~2726 degrees Centigrade or ~4940 degrees Fahrenheit).

However, if the universe didn't expand by a factor of 1000, and/or if the universe didn't expand at all, and/or if the universe was never at a temperature of 3000 degrees Kelvin (i.e., if the universe never "exploded"), and, consequently, if the universe didn't cool down during the process of expansion, and/or if thermal equilibrium between radiation and matter in intergalactic space were brought about in some other manner than in the way indicated by Big Bang advocates, then, one must seek some alternative explanation for the findings of Wilson and Penzias. The credibility of the Big Bang scenario depends on a great many conditions being true, and if any of those conditions are problematic (such as the issue of expansion that was brought into question in the previous chapter or considerations concerning conditions at the time of the Big Bang that were brought into question in the first chapter), then, one tends to be forced to revisit the puzzle concerning the meaning or significance of the data of Wilson and Penzias.

The data of Wilson and Penzias constitute facts. The interpretation of that data is another matter.

For example, the radio noise detected by Wilson and Penzias appeared to be fairly uniform (that is isotropic) regardless of the direction in which their radio antenna was facing. Many people felt that the foregoing sort of uniformity in radio noise is fully consistent with the idea that the Big Bang occurred in space everywhere at the same time, and, as well, such isotropic properties were considered by Big Bang advocates to be fully consistent with the idea that objects/particles were supposedly rushing away from one another as a result of the foregoing sort of explosive beginning, and, therefore, the Big Bang led -- or, so, proponents believed -- to a uniform smoothing out of the texture of the universe during the ensuing unfolding/expanding of the universe.

However, the foregoing sort of uniformity in radio noise is also consistent with another possibility. More specifically, the temperature of Interstellar space at thermal equilibrium in a steady-state universe would also tend to be uniform in nature and, as a result, would give rise to radio noise that would appear to be roughly the same irrespective of the direction in which one looked.

In addition, there is considerable evidence to indicate that the universe is not isotropic or uniform in structure ... at least not in any manner that is consistent with the way in which proponents of the Big Bang claim that things began in the universe. In other words, if the framework being advanced by proponents of the Big Bang were correct, then, one should be able to link what is observed today with how things supposedly began at the time of the Big Bang, but there are some significant problems in this respect.

According to proponents of the Big Bang, the universe began approximately 13.7 billion years ago. This figure is difficult to reconcile with some of the discoveries in astronomy that have been made since the mid-1980s.

For example, in 1986, astronomer Brent Tully indicated that he had detected the presence of a supercluster (consisting of an amalgamation of many galactic clusters strung together) that was 300 million light years in length and breadth, as well as 100 million light years thick. Given the speed with which many galaxies travel (e.g.,

Andromeda and the Milky Way galaxy are heading toward one another at about 130 kilometers per second), Tully estimates that the supercluster he discovered would have required approximately 80 billion years for it to have been able to form if it were operating in accordance with the scenario proposed by Big Bang advocates.

In 1989, a group of astronomers headed by John Huchra and Margaret Geller, reported detecting a structure that was referred to as: "The Great Wall". It consisted of a conglomeration of galaxies that formed a cosmic wall some 500 million light years long, 200 million light years wide, and 15 million light years deep.

Calculations indicate that such a structure would have required roughly 100 billion years to form if it were to have come together in the manner in which the Big Bang scenario claims things happened. This is more than 7 times the length of time that Big Bang proponents claim the universe has been in existence.

A group of American, Hungarian, and British astronomical researchers have reported the existence of structures that are even larger than the aforementioned 'Great Wall'. The international research group discovered a network of structures consisting of more than a dozen walls some 326 million light years long that are separated by voids that are roughly 600 million light years wide.

The time required for such astronomical structures to form in accordance with the conditions of the Big Bang has been estimated to be approximately 150 billion years. This is more than 11 times the period that – according to Big Bang proponents – the universe has been in existence.

In 2005, the "Sloan Great Wall" was discovered. It has been calculated to be about 1.36 billion light years in length, nearly twice as long as the 'Great Wall' discovered by the research team led by Huchra and Geller in 1989.

If one were complying with the conditions that supposedly operated in conjunction with the Big Bang, the "Great Sloan Wall" would have taken 250 billion years to assemble. This is nearly 20 times greater than the Big Bang's current estimate of the length of the universe.

One way of trying to reconcile a Big Bang scenario with the supercluster/void magnitudes mentioned in the last page, or so, would be to suppose that early on perhaps the speeds of galaxies were much faster than seems to be the case today and, that, at a certain point, those speeds must have slowed down. How and why the speed of galaxies would have been so high in the early universe but slowed down later on is unknown ... if this is even the way that events in the universe transpired.

Another approach to the problems that emerge in relation to current estimates of the age of the universe in the light of the huge size of various superclusters and voids that have been discovered raises the possibility that there might be something problematic with the way in which proponents of the Big Bang claim the universe unfolded. For example, maybe, the universe is much older than it has been calculated to be by some astronomers, but if the universe is older than it is presently estimated to be, then, assuming that a Big Bang actually occurred, the properties of a Big Bang theory will have to be altered in ways that can render it consistent with what had been discovered in conjunction with supercluster walls and huge spatial voids.

The problem involving current estimates concerning the age of the universe and the discovery of huge 'Walls'/'Voids' that has been outlined previously, does not end the challenges for Big Bang scenarios. There is at least one other problem that generates anomalies when one tries to reconcile the Big Bang assumption that the distribution of energy and material in the universe is fairly isotropic and homogenous (the Cosmological Principle) with the aforementioned discoveries concerning the existence of huge supercluster 'Walls' and 'Voids' in the universe.

More specifically, the Cosmological Principle holds that if one looks at the universe on a sufficiently large enough scale, then, everything will be seen to be distributed in a fairly uniform (isotropic) and homogenous way. If true, this means that the aforementioned 'Walls' and 'Voids' that have been discovered are part of a uniform and homogenous distribution of matter and energy in the universe, and if those 'Walls' and 'Voids' are partial expressions of an overall uniform and homogenous universe, then, the size of the universe would have to be sufficiently large to make such 'Walls' and 'Voids' relatively normal fluctuations in the overall scheme of things ... the sort of "normal" fluctuations that could be observed throughout the universe and that presumably are present in a uniform and homogeneous fashion.

A universe that is big enough to contain 'Walls' and 'Voids' that are hundreds of millions of light years long/wide/thick while, simultaneously, complying with the Cosmological Principle (and which would require the foregoing sorts of 'Walls' and 'Voids' to be uniformly and homogeneously distributed throughout the universe) is beyond the capacity of astronomy to observe. However, if we cannot see what the structure of the universe actually looks like, then, the Cosmological Principle is an idea that cannot be proven to be either true or false, and, as such, it is a philosophical notion, not a scientific one.

The universe that <u>can</u> be observed by astronomers today departs considerably from the universe that is presumptively described through the filters of the Cosmological Principle. The universe that <u>can</u> be seen is not uniform or homogeneous with respect to the manner in which matter and energy are distributed throughout it, and the great 'Walls' and 'Voids' are just one of the indications that we live in a universe that, in many respects (but not necessarily in all aspects) is anisotropic (exhibiting significant variations when measured in different directions) rather than being isotropic.

'Walls' and 'Voids' come in different sizes and shapes. Galaxies display different sizes and shapes. Stars have different sizes and properties. The components of various solar systems (e.g., our own solar system) exhibit different properties and sizes.

Differences rather than uniformity and homogeneity seem to characterize the universe. Of course, if one likes, one can average all of the foregoing differences away, but in doing so one filters reality through the artificial properties that are generated through the statistical manipulation of data rather than engaging reality as it is.

Statistics is a way of framing data. Sometimes that process of framing serves to bring certain themes into focus that might not otherwise be noticed, but, on other occasions, statistics gives expression to a framing process that distorts reality ... such as when someone says that the nuclear family in a given country consists of 4.2 people.

Proponents of the Big Bang claim that at a certain point the universe underwent a crucial transition with respect to radiation and matter. This threshold is estimated to have been traversed somewhere between 300,000 – 400,000 years following the Big Bang.

More specifically, when the temperature of the Universe lowered sufficiently to permit electrons and nuclei to be able to hook up with one another, the curtain of opaqueness that was being generated by the presence of ionized protons and electrons was removed. The removal of that ionized curtain of opaqueness supposedly provided the first clear glimpse of the Cosmic Microwave Background that would be observed by Wilson and Penzias in the form of radio noise that was emanated by a cooled down, isotropic remnant nearly 14.5 billion years later.

Wilson and Penzias restricted their initial observations to the radio noise being received at 7.35 centimeters. This was largely because what they discovered at that wavelength was so unexpected and puzzling.

However, other researchers began to look at other wavelengths of microwave radiation. For example, Wilkinson and Roll -- who, along with Dicke, had written a letter to the *Astrophysical Journal* concerning the findings of Wilson and Penzias (noted earlier) – released the results of the research that Dicke had induced them to undertake prior to their collaborative venture in letter writing with Wilson and Penzias.

Wilkinson and Roll focused on a wavelength of 3.2 centimeters. They calculated the temperature equivalent of the radio noise received at that wavelength to be between 2.5 and 3.5 degrees Kelvin.

The two researchers discovered that the intensity of the radio noise at 3.2 centimeters was greater than what Wilson and Penzias had recorded in relation to 7.35 centimeters. More importantly, the greater intensity of the radio noise at 3.2 centimeters compared with the intensity of radio noise at 7.35 centimeters was in line with what one would predict if the radiation being described by, on the one hand, Wilson and Penzias, and, on the other hand, by Wilkinson and Roll, were radiating from a black body.

In other words, if one were to consider the universe as beginning to display the properties of a black body at around the 300-400,000 year marker following the Big Bang (as ionization diminished significantly), then one should observe certain levels of intensity associated with different wavelengths. More precisely, those levels should correspond with what would be predicted by Planck's formula for blackbody radiation.

Indeed, the findings of Wilson, Penzias, Wilkinson, and Roll were consistent with the idea that Cosmic Microwave Background Radiation tends to reflect how black bodies would behave under various conditions of temperature. In fact, the intensity of the aforementioned background radiation has been studied across an array of wavelengths running from 0.33 centimeters up to 73.5 centimeters, and on each occasion, the results are consistent with the intensity of energy that a black body would radiate at different wavelengths when exhibiting a temperature between 2.7 and 3 degrees Kelvin.

Determining whether, or not, energy density will continue to fall off with decreasing wavelengths below 0.3 centimeters -- in accordance with Planck's formula for black body radiation involving temperatures between 2.7 and 3 degrees Kelvin -- is difficult to accomplish by means of radio antennas located on Earth. While the Earth's atmosphere is fairly transparent to wavelengths above 0.3 centimeters, the atmosphere tends to become opaque with respect to wavelengths below 0.3 centimeters ... and this tendency tends to become more pronounced as the wavelengths of interest become shorter.

However, through various methods, a variety of studies have been conducted involving shorter wavelengths, and the results of those studies all appear to point in the same direction as the research done with wavelengths above 0.3 centimeters. More specifically, irrespective of wavelength, Cosmic Microwave Background Radiation gives expression to black body radiation with a temperature that is at, or around, 3 degrees Kelvin.

As interesting as all of the foregoing results might be, nevertheless, establishing that the Cosmic Microwave Background Radiation first observed by Wilson and Penzias manifests itself in accordance with the properties of black body radiation at temperatures around 3 degrees Kelvin does not prove that such a state of affairs was necessarily brought about by a Big Bang of some kind. Steady State theories concerning the nature of the universe also have room for scenarios in which intergalactic regions of space could arrive at conditions of thermal equilibrium that, in turn, would behave like black bodies.

The central issue is not whether Cosmic Microwave Background Radiation gives expression to black body behavior. The central issue is whether we have arrived at this point of CMBR via a Big Bang or through a Steady State Universe or in some other way, and, to date, we are not in the sort of epistemological/scientific position to be able to plausibly claim that the only way that Cosmic Microwave Radiation that is detected today – or back in the mid-sixties -- could behave in the way it does, or have the properties that it does, is if there had been a Big Bang.

One might note in passing that there is nothing in the idea of a Big Bang that induces one to predict that such an event necessarily must end with a universe that operates as a black body in a way that is identifiable as something that would be unique to Big Bangs. The laws of thermodynamics on their own tend naturally to lead in the direction of giving expression to a universe that operates as a black body at thermal equilibrium quite independently of whether, or not, the universe came into existence as a Big Bang.

Furthermore, there are some indications that a Big Bang scenario would not even necessarily have led to conditions that would be characterized by black body properties or behavior. For instance, at the 2nd Crisis in Cosmology conference held in 2008, Bernard Bligh gave a paper indicating that a Big Bang would have led to a 'smeared' spectrum rather than a spectrum that is characteristic of black bodies such as is evident in Cosmic Microwave Background Radiation.

If Bligh is right, then, the black body properties of Cosmic Microwave Background Radiation is not proof of the Big Bang, but, rather, constitutes data that is inconsistent with such a theory. According to Bligh, Big Bang theories call for a different kind of spectrum than that which arises in conjunction with black bodies.

Bligh's thermodynamic analysis might, or might not, be correct. If it is, then, this is merely another nail in the coffin of the Big Bang theory as currently conceived, and if Bligh's perspective turns out to be incorrect, this does nothing to undermine the importance of the fact that scientists are unable to determine if the black body characteristics of the Cosmic Microwave Background Radiation constitutes a cooled-down remnant of the Big Bang or merely gives expression to energy/materials in intergalactic space that are now in thermodynamic equilibrium.

Some scientists point to the phenomenon of quantum fluctuations in energy and temperature as being a possible telltale indicator through which to differentiate between the viability of Big Bang and Steady State theories of the Universe. In other words, according to some individuals, one can link an early, hot universe to the current structure of the universe by means of the quantum fluctuations that supposedly occurred during the Big Bang and its aftermath, and such fluctuations would have led to the sort of shifts in the distribution of energy and matter that, over the course of billions of years – along with the assistance of gravity – would have resulted in the structures of the universe (i.e., stars, galaxies, clusters, superclusters, Walls, Great Walls, and so on) that are observed today.

However, to whatever extent quantum fluctuations actually occur (and I will have more to say on this shortly), one might suppose that, over time and with the assistance of gravity, quantum fluctuations – if they occur in the manner in which quantum field theory says they do -- might be able to lead to the formation of the aforementioned sorts of structures in a Steady State Universe as well as in a Big Bang Universe. If fluctuations in temperature and energy distribution on the quantum level actually serve as the seeds for large cosmic structures millions or billions of years later, then the general physics ought to be the same irrespective of whether one is considering a Big Bang scenario or a Steady State version of things.

As indicated previously, Wilson and Penzias found that the radio noise they were receiving at 7.35 centimeters was isotropic or uniform. Supposedly, the Big Bang occurred in such a way that it led to an isotropic distribution of materials and energy in the early universe, and the radio noise discovered by Wilson and Penzias was a cooled down version of the foregoing state of affairs that prevailed in the early universe.

Proponents of the Big Bang claim that temperature fluctuations of as little as one-millionth of a degree in the Cosmic Microwave Background are a reflection of density differences in energy distribution during the early universe. Moreover, as indicated earlier, many astronomers and physicists believe that such differences – along with many others -- are a function of quantum fluctuations that occurred in the early universe.

According to quantum field theory, virtual particles are blinking on and off in the vacuum of space. This process of virtual particles blinking into and out of existence allegedly gives rise to fluctuations involving temperature and energy density.

Quantum field theory also predicts that the total energy of the vacuum is extremely high, if not infinite. Yet, experiments that have been carried out with respect to measuring the energy density that is present in the vacuum of space point in a much more finite and modest direction.

No one actually knows to what extent quantum fluctuations take place in the vacuum of space … if they occur at all. Furthermore, if such fluctuations do occur, no one knows exactly what occurs during those fluctuations.

The notions of virtual particles and quantum fluctuations are a theoretical way of trying to account for certain aspects of what is observed. While the idea of virtual particles might help to make sense of various phenomena that have been observed, nevertheless, it is quite possible that the phenomena being described through the properties of virtual particles are actually an expression of some other kind of activity for which the idea of virtual particles merely constitutes an interim, descriptive placeholder … that is, something that helps makes sense of things for now but does not necessarily give expression to, or capture the nature of, what is actually transpiring.

The issue of quantum fluctuations is an interpretive artifact associated with the mathematics of the wave function. The wave function gives expression to a probability distribution, but no one knows why such a probability distribution has the properties it does or knows what induces (causes?) one dimension of that probability distribution, rather than some other dimension of that distribution, to manifest itself at any given time.

What is known is that to whatever extent fluctuations do occur in the vacuum of space, they do so (according to actual empirical data) in a fairly limited and finite fashion. Furthermore, it is quite possible that no quantum fluctuations whatsoever transpire in the vacuum of space and that the fluctuations in temperature and energy which are observed are due to a dynamic that, to varying degrees, can be statistically described (e.g., through the wave function) but involves phenomena that are beyond current understanding or beyond our present ability to follow on a quantum event by quantum event basis.

Did the allegedly uniform and homogenous conditions of the early universe contain the sort of fluctuations in matter and energy density that over time would have, on the one hand, led to the great 'Walls' and 'Voids' that are observed today, while, on the other hand, simultaneously generated the isotropic Cosmic Microwave Background that Wilson and Penzias observed in 1964? How does one reconcile the foregoing disparities in energy/matter distribution at the micro and macro levels that, supposedly, are entailed by one and the same Cosmological Principle?

If one assumes that fluctuations in temperature and energy in the early universe were a manifestation of the activity of virtual particles, then, statistically speaking, one might be able to scale those hypothetical fluctuations in a manner that is able to give expression to models that reflect certain aspects of the large-scale structure of the universe that exists today. Those models, however, are products of an underlying theory of virtual particles that have the capacity to fluctuate in ways that are capable of generating a continuous array of distribution patterns (in accordance with the mathematical potential of a wave function).

Yet, the available evidence indicates that fluctuations at the quantum level in the vacuum of space are fairly limited (i.e., finite) in the way in which they manifest themselves. Therefore, not every possibility inherent in the mathematics of a wave function will necessarily be realized ontologically.

In other words, given certain assumptions concerning the nature of virtual particles and quantum fluctuations, one might be able to generate a statistical analysis of Cosmic Microwave Background Radiation that alludes to a potential inherent in the structural variances on a micro level that might be capable of being seamlessly scaled up to the macro scale structures in the universe that are observed today. On the other hand, empirical evidence concerning the actual (rather than theoretical) nature of energy and material density distributions on the micro level indicates that the alleged potential of virtual particles and quantum fluctuations on the micro level might not be capable of being realized as great "Walls" and "Voids" on the macro level ... even if a theory says this is possible.

The wave function is a descriptive tool. It does not necessarily give expression to the nature of ontology other than to the extent that such a tool is capable of accurately describing a limited aspect of certain kinds of phenomena.

Indeed, when conceptual push comes to empirical shove, no one really knows what the precise nature of the relationship is between the mathematics of the wave function and the nature of reality. At the present time, the most anyone can say is that under certain conditions, wave functions are able to generate a set of mathematical possibilities that captures (i.e., predicts) how certain particles might behave in conjunction with specified conditions of: Energy, various laws of conservation, issues of thermodynamic stability, and so on.

The wave function intersects reality at certain points. We know the possible points of intersection <u>before</u> the fact by means of the wave function, but we don't know the actual point of intersection until some dynamic (such as a measurement) has taken place that identifies which probability possibility has been manifested due to dimensions of reality that seem to be beyond the pay grade of the wave function. (Certain aspects of the foregoing several pages of critical reflections are pursued in a little more detail within the commentary that takes place in 'Chapter 7: Physical Conundrums' of this book).

None of the foregoing considerations indicate that treating quantum fluctuations as either expressions of virtual particle activity and/or as a manifestation of the wave function is more conducive to a Big Bang scenario than to a Steady State account. Furthermore, none of the foregoing considerations demonstrate -- and quite independently of Big Bang and Steady State scenarios – that the sort of quantum fluctuations and virtual particle activities that supposedly transpire on the micro level constitute anything more than a theoretical account of possibilities that tend to run contrary (by many, many orders of magnitude) to actual measures of energy/matter density distribution in the vacuum of space.

Theory is one thing. Nonetheless, reality often turns out to be quite different than theory.

Two sources of data that allegedly demonstrate the veracity of the Big Bang theory are the satellite projects known as: COBE (Cosmic Background Explorer) and WMAP (Wilkinson Microwave Anisotropy Probe). COBE was placed in a sun-synchronous orbit on November 18, 1989, and WMAP was launched on June 30, 2001.

WMAP was intended to capture data that would serve as the means through which researchers would provide a more refined and detailed analysis of Cosmic Microwave Background Radiation than had been possible through COBE. COBE supposedly uncovered evidence that the primordial seeds of the large-scale structure of the universe were present in the Cosmic Microwave Background Radiation, while the data generated through WMAP allegedly not only confirmed -- in greater depth and with greater precision -- the findings of COBE but, as well, helped shape the understanding of scientists concerning the nature of the universe in a number of other ways as well (some of these other ways will be explored later in this book).

Both COBE and WMAP gathered evidence showing, among other things, that there was a dimension of anisotropy inherent in the Cosmic Microwave Background Radiation. In other words, although the scientists who developed and analyzed the data from COBE and WMAP believed that the Cosmological Principle correctly described the overall structural character of the universe in the sense that when the latter was viewed on a sufficiently large enough scale, then the universe would be seen to be isotropic (uniform) and homogeneous, nonetheless, the data generated through both COBE and WMAP were heralded because such data supposedly demonstrated that at its most basic level the universe was anisotropic. Moreover, the COBE/WMAP data supposedly demonstrated that when the universe is engaged on the most fundamental of levels differences are found from place to place in the distribution of energy being displayed in the Cosmic Microwave Background Radiation that, over time, scientists believed would generate the large-scale structure of the universe.

Somehow, the universe, in general, was supposed to be isotropic but, in specific terms, it was described as being anisotropic. This sounds comparable to the process of squaring the circle.

On both a micro level as well as on a macro level, the empirical evidence indicates that the universe operates in an anisotropic manner. Yet, many scientists keep insisting that the Cosmological Principle governs the universe ... that no matter in which direction one looks or how one measures the universe, nevertheless, the results will be isotropic (uniform) and homogeneous.

The reason for holding on to a principle that appears to run contrary to empirical data is because the Cosmological Principle is intimately tied to the belief that the Big Bang took place in such a way that energy and materials were scattered in a uniform and homogenous manner. However, if things were anisotropic from the very beginning, then, the dynamics of the Big Bang become a much more complicated process, and the truth is that proponents of even the simplest version of a Big Bang scenario (I.e., one operating in accordance with the Cosmological Principle) are unable to explain how the Big Bang took place or what would have led to such an event or why that event would have given expression to a supposedly isotropic and homogeneous universe rather than a universe characterized by anisotropic properties.

Another reason for holding tightly to the Cosmological Principle is that it fits in nicely with the findings of Wilson and Penzias in relation to the presence of a radio noise at 7.35 centimeters that seems to be isotropic and homogenous no matter in which direction it is measured. As such, the data from the research of Wilson and Penzias appears to serve as a confirmation of the truth of the Cosmological Principle, and, therefore, in turn, serves as a confirmation of the Big Bang theory ... a theory that assumes the primordial explosion took place in an isotropic and homogenous manner.

Nevertheless, the findings of the COBE and WMAP projects indicate that the microstructure of the Cosmic Microwave Background Radiation is not uniform -- rather, it is anisotropic. Consequently, whatever the significance is of the anisotropic character of the microstructure of the radio noise that is referred to as Cosmic

Microwave Background Radiation, the presence of the foregoing sort of anisotropy undermines the Cosmological Principle.

In addition, given that the Big Bang supposedly unfolded in accordance with the Cosmological Principle, the Big Bang model did not predict (ore even suspect) that the property of anisotropy would be found to be an essential feature of the Big Bang model. Therefore, as far as the issue of anisotropy is concerned, the findings of COBE and WMAP do not necessarily constitute a confirmation of the Big Bang theory ... at least not as originally conceived.

Of course, someone might try to argue that, for the most part, the universe is really isotropic and homogeneous in nature, but it has just enough of the right kind of anisotropic features at a quantum level to be able to produce large-scale structural differences in an otherwise isotropic universe. Quite frankly, this sounds more like Orwellian New Speak than it sounds like science.

Looking at the universe through the filters of the "right" scale of magnitude, the universe appears to be homogeneous. However, when one looks at the universe through the filters of an appropriate scale of microstructure, the universe no longer appears homogeneous, and to try to claim that the former scale of magnitude is the way in which we should look at the universe seems, at the very least, quite arbitrary.

On the basis of available evidence, jettisoning the Cosmological Principle would appear to be a prudent thing to do. However, this doesn't mean one must also throw out the idea of some kind of Big Bang event.

Nonetheless, relinquishing ties with the Cosmological Principle does mean that if one wishes to hold on to a Big Bang perspective, one is going to have to refurbish that theory in a number of ways. Such remodeling could start with the idea that the Big Bang might have been inherently anisotropic from the very beginning (the relationship between matter and antimatter that was discussed in Chapter 2 of *Final Jeopardy: Physics and the Reality Problem, Volume II* might serve as another indication that the universe was anisotropic from the get-go).

The foregoing analysis has been fairly simple. There are other, more technical and complicated analyses that are available and which go into considerable technical detail concerning the possible problems entailed by the COBE and WMAP projects.

Those analyses call into question whether, or not, the COBE and WMAP data actually demonstrate what the scientists associated with those projects claim they do. Some of these critiques point out problems with the way in which the COBE and WMAP instrumentation was designed or used, as well as raise questions about how the data generated through those instrument packages were interpreted.

For example, Pierre-Marie Robitaille became engaged in a radiological analysis of the instrumentation used in COBE and questioned whether the design features of that instrumentation could lead to the sort of data that are capable of reliably substantiating the conclusions that project scientists were claiming in conjunction with that instrumentation. Among other things, Robitaille indicated there were design flaws in some of the instrumentation on board COBE (e.g., FIRAS, Far-Infrared Absolute Spectrophotometer) that would have led to misleading and erroneous conclusions concerning the data generated through those instruments.

Robitaille raised similar questions involving design flaws with respect to some of the instrumentation included in the WMAP project. In fact, Robitaille went so far as to say that some of those instruments might be generating anisotropic data that was not originally present in the Cosmic Microwave Background Radiation being sampled.

Robitaille's analysis is far too technical to be discussed in any detail here. To properly explain his position would require more space to explain than I am prepared to donate.

However, the fact that a well-respected scientist is raising questions about the integrity of instrument design and, as a result, is raising questions about whether such instruments will permit anyone to draw reliable conclusions concerning the meaning of the data being generated through those instruments, suggests, at the very least, that one ought not to accept anyone's claims at face value. No matter which side of things one is considering, just because various pronouncements of one sort, or another, are being voice by someone who is referred to as a scientist and who works with high-tech equipment is not an adequate basis for accepting what is said.

Knowledge is not about what someone else says or having an opinion about what is said. Knowledge is about the quality of the understanding one develops while exploring a given area.

It might be the case that Robitaille's analysis is what is flawed rather than the design of the instrumentation on which he is critically reflecting. Nonetheless, irrespective of whether Robitaille is correct or not, several problematic issues linger on in relation to COBE, WMAP, and Cosmic Microwave Background Radiation.

Firstly, as noted previously, the presence of anisotropic elements in the data from COBE and WMAP does not necessarily constitute a confirmation of the Big Bang theory. This is especially noteworthy given that such a theory – at least, as initially conceived – was rooted in a Cosmological Principle that claimed the universe was isotropic and homogenous while the evidence from COBE and WMAP indicate otherwise.

Secondly, the existence of Cosmic Microwave Background Radiation cannot be uniquely tied to the notion of a Big Bang in a way that persuasively eliminates the possibility that such radiation might only be giving expression to what happens at thermal equilibrium in intergalactic space. In other words, both Big Bang theories as well as Steady State models have ways of accounting for the existence of such background radiation, and, therefore, the existence of that sort of radiation does not, in and of itself, constitute a smoking gun that points definitively in the direction of Big Bang theories.

Thirdly, whatever anisotropic characteristics have been found in the data generated through COBE and WMAP, one cannot automatically conclude that such anisotropic elements are remnants of what was taking place when ionized nuclei and electrons were being decoupled from the rest of the radiation that was present in a supposedly hot, early universe. Conceivably – and, perhaps, even more plausibly – such anisotropic elements could merely be giving expression to the dynamics that take place at thermal equilibrium.

To whatever extent quantum fluctuations and virtual particle activity take place within Cosmic Microwave Background Radiation, one might anticipate the presence of a certain degree of anisotropy no matter how such radiation arose (e.g., the Big Bang or the thermal equilibrium of a Stead State universe). Therefore, one cannot necessarily conclude that the presence of anisotropic properties in background radiation gives expression to the cosmic "seeds" from which the large-scale structure of the universe would emerge billions of years later.

The anisotropic properties that emerge from place to place at thermal equilibrium are just anomalies that arise and, then, disappear. They are anomalies that do not appreciably affect the future state of things because they are damped down and balanced out by the presence of other anomalous perturbations of a lesser or greater nature.

Proponents of the Big Bang interpret the foregoing sorts of anomalies as "signs" of things to come. However, like the tealeaves at the bottom of a fortuneteller's cup, what can be read into such signs might be more an indication of the reader's imagination than it is of anything else.

One of the discoveries arising from WMAP data involves evidence allegedly confirming that the universe experienced an intense period of inflationary expansion. This process of inflation supposedly permitted the universe to expand a trillion, trillion times in a trillionth of a trillionth of a second, and during this time, small fluctuations occurred that formed the seeds from which galaxies would later arise.

The fluctuation issue already has been explored to some degree. Let's take a look at the issue of inflation.

Prior to the time in the early 1980s when Alan Guth introduced his inflationary model of the universe, the Big Bang theory had run into some difficulties. Among these were the flatness, horizon and monopole problems.

The Big Bang theory is tied to initial conditions that need to be fine-tuned to certain values if one is to be able to derive the universe we see today from the universe that existed during the early, hot universe. If such values differed by even a small amount in the beginning, the current universe would not have the properties it does.

For example, the present energy/matter density of the universe appears to be fairly flat. In the parlance of General Relativity, the density of energy/matter in the universe is such that its impact on the curvature of space-time is virtually negligible.

Over time, energy/matter density would tend to deviate fairly quickly from the value that is necessary for a flat universe. Consequently, physicists have calculated that the energy/matter density of the early universe must have been even closer to the critical value needed for flat space-time than is the case today ... perhaps differing from that value by as little as one part in $10^{62}$.

Robert Dicke pointed out the nature of the flatness problem in 1969. How and why was the density value of the early universe so finely tuned to the critical value that gives expression to the flat universe that is observed today?

Around the time Dicke raised the foregoing question, Charles Misner introduced another problem concerning the Big Bang theory. This was known as the "horizon problem".

More specifically, if one treats the known universe as a circle, and the circumference of that circle is said to be the horizon of what can be detected by us, then, points on the horizons that are opposite one another are too far apart to be able to communicate with each other through any sort of light-based means that is consistent with the theory of special relativity as currently understood. Yet, as far as we have been able to determine, the horizons on each side of the universe are characterized by the same kinds of temperature and set of physical properties as one another, so if, the distances between points on the horizon are too far apart to be able to exchange information, then how does one account for the similar values of temperature (such as in the Cosmic Microwave Background Radiation) and various physical properties that are found throughout the known universe?

Finally, according to some theorists, Maxwell's equations alluded to the possible existence of monopoles. Calculations have been in conjunction with the extremely high temperatures that, supposedly, were in existence at the time of the Big Bang, and according to such calculations, there should be many monopoles in existence.

Monopoles, if they exist, consist of one magnetic pole (either a south or a north pole without its counterpart). Therefore, they have a net magnetic charge.

Monopole particles are predicted by both superstring theories as well as grand unified theories. However, although certain kinds of condensed matter contexts appear to give expression to a quasi-form of monopoles and to monopole-like phenomena, no one has ever observed a monopole particle, and Guth hypothesized that the reason for this is that an inflationary event diluted the concentration of monopoles and turned them into hard-to-find, rare entities?

Guth came up with a theory that he felt might resolve all three of the foregoing issues (flatness, horizon, and monopole problem) in one fell swoop. More specifically, he hypothesized that in the early universe (around $10^{-35}$ seconds) there was a very brief (lasting, perhaps, between $10^{-33}$ to $10^{-32}$ seconds) but very rapid, period of cooling that occurred during a delayed phase transition such that a false vacuum (a temporary and unstable state of low energy/matter density) was created.

According to Guth, the false vacuum would rapidly decay as a result of quantum tunneling. Quantum tunneling is a term that refers to the possibility that a given particle has a finite probability of appearing (through unknown means) on the other side of a given energy barrier that normally cannot be crossed.

Through quantum tunneling, the aforementioned false vacuum would decay into a true vacuum. In the process, the universe would expand at a tremendous rate (by a factor of $10^{25}$ to $10^{30}$, or more) as a result of the negative pressure that was created and, therefore, was able to push things apart to such an extent that the volume of the universe was increased by a factor of at least $10^{78}$.

What the nature of the aforementioned phase transition was or why it was delayed is not known. What turned the process of inflation on, or what turned it off, or why it lasted for the length of time it did, is not known ... although such questions presumably have something to do with the creation and decay of a false vacuum of some kind.

Irrespective of whatever problems are allegedly resolved by virtue of Guth's notion of cosmic inflation, that theory becomes realistically plausible only if the blanks concerning the nature of the unknown features noted above are filled in with the sort of specific, demonstrable, empirical data that is capable of underwriting the claims that are being made. Otherwise, Guth's theoretical position merely becomes like someone going crazy with a credit card without worrying about the fact that at some point the benefits one is enjoying through the charging process will face the ugly reality that payments of a concrete sort will have to be paid with monetary denominations that are worth something [(i.e., Real data that justify the charging (hypothesizing process) are needed to properly finance the things one is enjoying in the short run (e.g., cosmological problems that are resolved].

The idea of symmetry breaking might provide answers to the foregoing unknowns. Various kinds of symmetry breaking supposedly occurred in conjunction with the appearance of the Higgs field and/or the unfolding of the electro-weak theory as a single force allegedly differentiated into the electromagnetic and the weak forces during an early stage of the Big Bang. However, the precise character of such symmetry breaking isn't any clearer than is the nature of the idea of a delayed phase transition that supposedly leads to inflation.

Furthermore, the notion of quantum tunneling that forms part of the idea of cosmic inflation doesn't really explain anything. As touched upon earlier, quantum tunneling merely alludes to the idea that there are calculations that can be made indicating that a given particle is likely to end up on one side of an energy barrier rather than the other side, but how such possibilities are translated into a reality of a certain kind is unknown.

For many physicists, the alluring quality of inflation theory is its alleged capacity to resolve a number of problems confronting the Big Bang theory (e.g., the horizon, flatness, and monopole problems). The downside of cosmic inflation is that it merely replaces one set of unknowns (the aforementioned problems) with, yet, another set of unknowns (the details of the actual process through which cosmic inflation took place ... if it did).

If one assumes that certain things happened in the early universe (such as the switching on and off of the inflationary process), then cosmic inflation tends to make a certain amount of sense. On the other hand, if the events that are presumed to have happened did not actually occur, then, cosmic inflation appears to constitute little more than an arbitrary fudge factor intended to sweep certain problems under an explanatory carpet that covers up more than it reveals.

Without cosmic inflation, a number of important facets of the Big Bang theory become problematic. In order for the Big Bang scenario to work properly, then, seemingly, some version of cosmic inflation must be true.

However, so far, there is no independent evidence that has been uncovered demonstrating the existence of a mechanism in the early universe that would have been able to give expression to inflation in the manner proposed by Guth or by those who, over the years, have made various suggestions in an attempt to make the underlying idea of inflation more credible. Everything connected to cosmic inflation is of a highly speculative and hypothetical nature that relies on assumptions rather than hard data.

Interestingly enough, at least one prominent scientist – Roger Penrose -- has put forth a statistical argument indicating that given an array of starting configurations involving possible inflationary and gravitational fields, a flat universe is more likely to arise through non-inflationary conditions rather than through inflationary ones. Penrose calculates the differential likelihood between non-inflationary and inflationary possibilities to be around $10^{100}$ in favor of the former venues.

Other researchers also have arrived at conclusions similar to those of Penrose on this matter. For example, using an extrapolative approach that works out the physics of various possibilities running backward from the present universe to early starting points, Neil Turok and Gary Gibbons have calculated that many of their extrapolations (the predominant number) involve insignificant amounts of inflation and, yet, still arrive at the sort of smooth, flat universe that we view today.

If cosmic inflation is not needed to resolve the flatness issue, then, is Guth's theory as attractive as it seemed to be initially? Or, consider the monopole issue ... another problem that, supposedly, becomes resolved – at least to a degree -- through the process of cosmic inflation.

The story goes as follows: Perhaps the reason why no one has detected monopoles – despite the fact that they have been predicted to exist – is because cosmic inflation has helped to render something that is rare even rarer. More specifically, calculations have been performed indicating there is less than one monopole for every $10^{29}$ neutrons or protons, so even under the best of circumstances, such entities tend to be very scarce, and cosmic inflation might have made them even scarcer by separating them from one another – and the rest of us – through the huge increase in volume (by a factor of $10^{78}$) that has been theorized to occur during cosmic inflation.

One can, of course, resolve the missing monopoles in another way that does not depend on the process of cosmic inflation. Perhaps monopoles don't exist at all because the temperatures needed for their formation never existed, and if this is the case, then one doesn't need cosmic inflation to explain why monopoles have not been found.

Consequently, some of the advantages that supposedly come with cosmic inflation might not be as valuable as once seemed to be the case. On the other hand, the problems entailed by the dynamics of cosmic inflation continue to persist.

For example, central to Guth's perspective is the idea that space is something that is capable of inflating. But, is space inflatable?

We don't even know what the nature of space is. So, how can we know whether, or not, it is inflatable?

Does space have a structure of some kind, or is it without structure? Later on in this book, some of the theories concerning the possible structure of space will be explored, but irrespective of whether, or not, space has some sort of structure, the existence of such structure does not automatically render it susceptible to inflation.

If space is not 'something' that is inflatable, then, Guth's whole theory falls apart. After all, one of the most intriguing features of Guth's idea concerning cosmic inflation is that, intuitively, one can see how if space expanded, then the objects occupying space wouldn't have to move in order to be pushed apart  ... instead, the space between particles would increase and this would have the same effect as if the particles, themselves, had moved.

Initially, some individuals criticized Guth's perspective because they thought that during the process of inflation particles would have to be moving at superluminal speeds (in violation of the principle of special relativity) in order to achieve the sort of increase in volume that Guth theorized had taken place in the early universe. However, because space is what was expanding -- and not the particles occupying space -- the foregoing criticism could be dismissed.

On the other hand, if space does have an inflatable structure, then its properties must be rather amazing. Indeed, one wonders how structural features of any kind are able to inflate (and, therefore, undergo transitions and transformations) at the rates required by Guth's calculations.

The idea that space has some set of structural properties that have the capacity to inflate stands in need of both clarification and an explanation. Unfortunately, the nature of the physical dynamics that would be capable of inflating space in the way Guth envisions is, at the present time, unknown.

In light of the unresolved questions currently surrounding the issue of spatial structure, one might wonder if it is possible that the expansion of some sort of scalar field (a quantity having magnitude but not direction) that is independent of space could bring about the results Guth has calculated? I think the short answer is: "No!"

Guth's cosmic inflation theory needs the spatial volume of the universe to increase without affecting or interacting with the ions that populate the universe at the time of inflation. While there might be an increase in volume of a scalar field if that field expanded, nonetheless, particles would remain in relatively the same spatial location as they were prior to the period of expansion.

If particles were affected by the inflationary dynamic of a scalar field (that was independent of space) and, as a result, those particles got moved along by the inflationary force of the scalar field, then, one would encounter a whole set of problems requiring explanation. Such problems likely would include the need to explain how that kind of a scalar field interacts with and, therefore, affects matter, and, therefore, such problems also would entail the need to provide a way of resolving the superluminal problem touched upon, and dismissed, previously.

Guth's theory of cosmic inflation is attractive because it bypasses all of the foregoing issues ... although in doing so, the theory generates a number of other problems – previously noted – that neither Guth nor anyone else has resolved satisfactorily. However, if space is not inflatable, then, one has no plausible way to increase the volume of the universe by a factor of $10^{78}$ within $10^{-33}$ and $10^{-32}$ seconds without entailing a number of seemingly insurmountable problems in the process.

Since Guth first proposed his idea of cosmic inflation, the idea has undergone some renovations. For example, Alfred Linde maintains that once inflation begins it will not stop.

According to Linde, quantum fluctuations are an inherent feature of inflation and such fluctuations will delay the point at which inflation comes to an end. Since the nature of those fluctuations are considered to be random in character, then in certain regions of space one will encounter only small delays ensuing from such random fluctuations, while in other regions of space, those random fluctuations will lead to substantial delays in the termination of the inflation process.

The impact of random, quantum fluctuations on the inflationary process is considerable. One will not only tend to have regions of space that continue to inflate surrounding islands of hot matter and radiation, but, as well, the regions of space that are continuing to inflate will also generate further regions of space in which islands of hot matter and radiation are surrounded by regions of space that are continuing to inflate and producing further islands of hot matter and radiation surrounded by additional regions of spatial inflation ... and so on.

Linde believes that the aforementioned islands of matter and radiation will not all be the same. They will differ due to the random nature of the quantum fluctuations that is supposedly driving the process of inflation.

Some number of those islands will be similar to the portions of the universe that are visible to us ... in fact, according to Linde, there will be an infinite number of such islands. However, there also will be other islands of matter and radiation that will give expression to conditions with which we are not currently familiar, and these other islands also will be infinite in number.

Does quantum reality continuously fluctuate? If it does, are such fluctuations necessarily random?

Measurement can only capture part of what is taking place. What occurs beyond the horizons of a given measurement is unknown.

The wave function doesn't indicate that an array of possibilities beyond whatever is being measured are also occurring. The wave function indicates that whatever happens will take place within the frame of reference indicated through the wave function.

The wave function is like placing an opaque, template screen over reality. The screen has holes in it that correspond to the likelihood that a certain kind of event (of given mass, charge, spin, and so on) will manifest itself through one, or another, of the holes assigned to the screen by the wave function.

Nonetheless, the screen template does not constitute anything but a relational reality ... a relational reality in which the methodology frames our engagement of reality in accordance with the properties of that methodology. That methodology constitutes a prediction concerning where, when, and how certain kinds of events are likely to show up ... nothing more.

Is anything happening outside of the holes on the template that is generated through the wave function? We don't know.

Is anything happening at any of the template holes that are assigned through the wave function that do not turn out to be the hole that gives expression to the value for a given measurement or physical event? We do not know?

Treating the template generated through a given wave function as if it were the reality that is being measured is a mistake. The methodology (in this case, the wave function template) and the ontology being measured are two different manifestations of reality.

Moreover, one cannot necessarily conclude that the event being framed through the wave function template is random in nature. In fact, quantum mechanics indicates again and again that there is a measurable order to the way things happen, but although we can predict the likelihood that a given event will have one set of characteristics rather than some other set of characteristics, we just don't necessarily know what the nature of that manifested order will be ahead of time.

Not knowing what the outcome will be does not make something random. The foregoing epistemological condition merely means that our understanding of how things work is permeated by a considerable amount of ignorance.

Linde really has no knowledge that inflation will continue on indefinitely. He has no knowledge that quantum fluctuations are occurring beyond the horizons of a given measurement or observed event.

He has no knowledge that there are sets of simultaneous events that are occurring beyond the horizons of measurement or observation that are random in nature. Indeed, Linde has no knowledge that what is being measured or observed is a random event or that it will be infinite in nature.

Linde is using the notions of quantum fluctuations and randomness as if they are demonstrable ontological realities. But, they don't ... they give expression to a hypothesis about what Linde believes is happening beyond the horizons of whatever he is currently measuring or observing.

Even granting the idea of inflation, is one necessarily required to suppose that inflation is governed by ontological properties that generate a never-ending series of random, quantum fluctuations during such a process? The necessity that is inherent in Linde's eternal inflation proposal appears to be more a function of his underlying assumptions concerning randomness and the nature of the wave function than it is a accurate reflection of ontology's character.

Linde – along with many other physicists – is trying to make the wave function something that it is not. The wave function is not a reflection of reality but, rather, it is a predictive function concerning the properties of certain outcomes ... without any understanding of why such outcomes take place when and as they do.

To claim that such outcomes are random in nature is to make an assertion that is rooted in a lack of understanding about why things occur in the way they do. Randomness is not an explanation, but the absence of an explanation.

A claim of randomness indicates that one is unable to discern an underlying pattern in the occurrence of events. It does not say there is no such pattern present ... randomness is an assumption about the nature of reality.

Even if, for the sake of argument, one grants inflation and the idea of quantum fluctuations, there is nothing that requires one to suppose that either inflation or quantum fluctuations are ontologically random in nature (rather than being a function of a methodologically-based notion of randomness). Moreover, as indicated earlier, both the idea of inflation and the notion of quantum fluctuations are problematic assumptions concerning the nature of reality.

Some scientists believe that detecting the presence of gravitational waves will help to substantiate claims concerning the occurrence of an inflationary event of some kind during the early universe. However, as will be touched upon later in this book, there might be other ways of generating delicate, gravitational ripples in the Cosmic Microwave Background that are not inflationary in nature ... assuming, of course, that gravitational waves do occur and that they have been detected.

## Chapter 4: The Electric Universe

According to some scientists, 96% of the energy/matter of the universe does not consist of "stuff" with which we are, to some extent, familiar -- such as: Protons, neutrons, electrons, and photons. Apparently, dark matter and dark energy make up most of the physical universe, and, therefore, whatever understanding we have, is limited to just four percent of the available possibilities.

Exactly what dark matter and dark energy entail is uncertain. Several later chapters will explore those two topics to varying degrees, but for now, one might just raise the following issue: If we don't understand 96% of the material character of the universe, one wonders why cosmologists and other scientists seem to be so confident in their pronouncements concerning the nature of the universe and its origins.

The foregoing question becomes more critical when one discovers that even the 4% that, supposedly, is known has been framed by ideas that remove many possibilities from consideration. For example, some physicists believe that particles in a plasma state (highly conductive ionized gasses consisting of moving positive ions and negatively charged particles) occupy most of the universe and that such plasma states might have a far greater impact on shaping the character of the universe than most astronomers have been, and are, willing to concede.

Some physicists believe that plasma dynamics might have the capacity to produce the phenomena for which terms such as "dark matter" and "dark energy" have been coined. In other words, what many modern scientists consider to be new, strange, and mysterious dimensions of the universe might just be a function – at least in part -- of some of the possibilities that are inherent in plasma physics.

Many plasma physicists believe the nature of cosmology could be revolutionized if scientists started to pay attention to the extensive roles that plasma physics appears to play in the dynamics of the universe at large. Such roles extend from: The formation of stars, to: the structuring of galaxies, galaxy clusters, and planetary systems.

Early in the 20th century, Irving Langmuir, who later won a Nobel Prize in Chemistry, began to study the phenomenon of electrical discharges in low-pressures gasses. Borrowing an idea from biology, he applied the term "plasma" to the phenomena associated with the electrical discharges in gases as a way of alluding to some of the self-organizing, lifelike properties that often emerge in ionized clouds of particles that were subjected to the presence of magnetic fields and electric currents.

The same laws of physics that govern the activity of gases do not regulate plasmas. The ionized particles that constitute plasmas comply with the laws of electromagnetism.

Beyond the horizons of artificial, neon signs, plasmas are present in many natural contexts. For example, the ionosphere of Earth gives expression to the presence of plasma activity in the form of auroras, and lightning is another plasma-related phenomenon that occurs in conjunction with the Earth.

The sun's corona operates in accordance with plasma physics. Moreover, the solar winds that sweep out from the upper atmosphere of the Sun are another example of plasma activity.

The aforementioned solar winds help shape the heliosphere. This is the volume of plasma generated by the sun that envelopes the solar system and, among other things, tends to establish the magnetized boundaries that mark where the solar system ends and interstellar space begins ... a boundary that was penetrated by Voyager 1 on August 25, 2012.

Plasmas occur in three general states. These are known as: (1) Dark current mode, (2) normal glow mode, and (3) arc mode.

Low-strength electric currents characterize the dark current mode of plasmas. Such plasmas do not glow and tend to radiate longer-wavelength radio waves.

The electric currents in the second category of plasma noted earlier are, as the name suggests, sufficient to generate a glow throughout the plasma. The density of the plasma, as well as the intensity of the current running through it, will determine how brightly such plasmas will glow, while the color of that glow is a function of the nature of the ions present in the plasma.

Very high electric currents are usually present in the third kind of plasmas -- the arc mode. These currents often assume the form of twisting filaments that radiate across a wide spectrum of possibilities, including ultraviolet wavelengths.

The aforementioned Irving Langmuir discovered that plasmas have the capacity to form walls that separate off one region of plasma activity from other regions in the same plasma field. These walls are made of two, charged layers -- one is negative and one is positive – that are closely spaced relative to one another.

The foregoing charged walls are known as double layers (DL). The strongest electric fields in plasmas are associated with such double layers.

Langmuir also discovered that plasmas exhibit a frequency. This frequency is a function of the way that free electrons move to and fro with a harmonic motion relative to slower-moving, positively charged ions in the plasma.

In addition to the foregoing sorts of discoveries, scientists found that materials (whether ionized or unionized) can be compressed when caught between high-intensity electric currents that run through plasmas. This is known as the 'z-pinch' effect or the Bennett pinch (in honor of W. H. Bennett who first began studying the phenomenon in 1934).

For example, Birkeland currents are spiral-shaped forms that occur in plasmas characterized by high-intensity electric currents. Such currents tend to occur in pairs and have the capacity to compress or pinch materials that are lodged between them, and, as a result, some plasma physicists believe that many of the filamentary-like structures that are found in the cosmos might be the result of powerful pinching processes that are capable of occurring in plasmas.

The electromagnetic forces coursing through the plasmas of the cosmos are far more dominant than are the gravitational forces that exist there. This is so by 39 orders of magnitude, and the difference in strength between those two forces – electromagnetic and gravitational -- is exemplified by the fact that a small hand-held magnet is capable of lifting a metallic object into the air, and in doing so, a relatively tiny magnet effectively counteracts the entire gravitational force of the Earth.

Hannes Alfvén, an electrical engineer from Sweden, won the Nobel Prize in physics in 1970 for his research into magnetohydrodynamics. That work involved, among other things, describing a class of MHD waves that have come to be known as Alfvén waves.

Initially, plasmas were considered to be perfect conductors. Consequently, taking his lead from Maxwell's equations, Alfvén believed that when magnetic fields occurred in a context of a perfect conductor, those fields could not vary in any fashion and would appear to be fixed or frozen.

The aforementioned theory of magnetohydrodynamics focused – as the term suggests -- on the magnetic facet of electromagnetic phenomena. Consequently, physicists took the equations that had been derived to describe those dynamics (which had more to do with the flow of fluids than it did with electromagnetism) and applied those equations to plasma activity in the cosmos without necessarily taking into consideration the contributions made by the other half of those phenomena – that is, their electric dimension.

For more than three-fourths of the 20th century, astronomers and physicists believed that the vacuum of space was incapable of conducting electric currents because charge separation was considered to be impossible under such conditions. When scientists subsequently discovered that space is filled with a plasma that has the property of conductance, many physicists altered their position somewhat and proceeded to filter their understanding of plasmas through Alfvén's original description of magnetic fields ... that is, fields that were "frozen" forms of neutral plasma and, therefore, scientists believed that electric currents in the universe were a function of the way plasmas were magnetized and, thereby, became able to serve as conductors.

According to the foregoing scientists, we didn't live in an electromagnetic cosmos. Rather, we lived in a magnetic universe that could be induced to conduct electric currents under certain, limited conditions.

Thus, the default model of cosmology indicated that electric currents were not, and could not, ripple through the universe. Rather, those currents were considered to be localized, temporary phenomena that were capable of magnetizing plasma and, then, such electrical

| Cosmological Frontiers |

activity could be largely ignored when developing models of cosmic plasma dynamics.

Eleven years after receiving his Nobel Prize, Alfvén wrote a book entitled: *Cosmic Plasma*. The book took issue with the direction in which astrophysics and cosmology were headed due to the failure of the latter disciplines to take the electromagnetic character of plasma physics into account while formulating their theories of the universe.

The seeds for the foregoing book were actually present in the acceptance speech given by Alfvén during the 1970 Nobel ceremonies. At that time, Alfvén indicated that the research on magnetohydrodynamics for which he was being awarded recognition needed to be expanded in order to accommodate electricity. In other words, he believed that his Nobel research only presented half the picture, and scientists who left out the electric component of the forces operating through the plasmas of the universe would likely construct a distorted and incomplete, if not erroneous, understanding of the universe.

Alfvén's thinking began to change due to, among other things, the discovery that while plasmas are very good conductors, they are not perfect conductors. When filtered through Maxwell's equations, this new realization meant that one no longer had to consider magnetic fields to be frozen within plasmas, and, therefore, the newly acquired understanding left room for magnetic fields to be able to move within plasmas and, in the process, generate electric currents.

Unfortunately, mathematicians who had never spent much, or any, time in actual laboratories that empirically explored plasma dynamics were often the very individuals who were conjuring up mathematical models (often steeped in the flow of fluids) to describe those dynamics. Even more unfortunately, despite the fact that Alfvén and other researchers could provide evidence in their laboratories demonstrating that the foregoing sorts of mathematical models were incorrect, many theorists insisted on holding on to their beloved mathematical models concerning plasma dynamics.

One of the most important principles that Alfvén established through his post-Nobel work is that the magnetic fields occurring in the plasmas of space cannot be permanently frozen in the manner that his original work on magnetohydrodynamics had indicated. In space, plasmas require the presence of electric currents to both create and maintain those fields.

Back in 1942, Alfvén already had imagined a possible way of accounting for how charge separation might take place in space. More specifically, if an ionized cloud (i.e., plasma) were to travel with sufficient velocity through a neutral cloud of gas, the latter would become ionized.

He estimated that the requisite ionization velocity would be somewhere between 5 to 50 kilometers per second. Alfvén's hypothesis was confirmed in 1961 and has come to be known as Alfvén's critical ionization velocity.

Galaxies contain plasma fields. Given that galaxies travel through space at speeds of 100 kilometers, or more, per second, galaxies would be one way in which many of the neutral gas clouds in intergalactic space might become ionized plasmas ... plasmas that, in turn, could subsequently generate electromagnetic behavior of one kind or another.

Alfvén maintained that the plasma phenomena being explored in the lab were fully scalable to the cosmic level. In other words, the principles and properties governing the activity of plasmas in the laboratory were also present when one considered such phenomena on a cosmic scale.

For example, during the 1980s, astronomers were coming across more and evidence that a great deal of cosmic structure was filamentous or stringy in character. Such structures were entirely consistent with the phenomena that could be observed in Earth-bound, plasma laboratories ... albeit on a smaller scale.

According to Alfvén, cosmic plasmas generate networks of circuits that are capable of transferring energy over considerable distances. Those networks establish boundary areas consisting of a double-layer of filament-like properties that have been pinched off from the rest of the network and, therefore, tend to form cellular regions that are insulated from one another.

Alfvén discovered that incredibly strong electric fields often operate across the aforementioned boundaries. He believed that understanding the properties of those fields is crucial to understanding the behavior of plasmas.

The capacity of gravitational forces to shape the universe is important. However, the capacity of the electromagnetic properties operating within, and through, plasmas across the universe might be as cosmologically important, if not more so, than gravitational forces are.

Prior to the work of Alfvén and other plasma physicists, most scientists believed that the organizational work (e.g., in relation to the formation of stars, galaxies, clusters, and so on) that were observed taking place across the incredible distances of intergalactic space was predominantly due to the effects of gravitational activity. Following the trailblazing efforts of Birkeland, Langmuir, Bennett, Alfvén and others, some scientists began to entertain the possibility that the electromagnetic activity taking place in conjunction with the plasmas occupying intergalactic space might have the potential to accomplish considerable cosmological structuring as well.

Initially, the extent of such cosmic electromagnetic activity was difficult to detect since astronomers were restricted to what could be discovered through the use of light-telescopes operating in the visible spectrum. Of course bigger and better versions of these light-telescopes could be constructed at higher attitudes and/or in locations with more favorable atmospheric conditions, but such creations still were limited in various ways with respect to what they could reveal about what was talking place in the universe.

Later on, Karl Jansky, an engineer, uncovered – quite by accident -- the existence of radio waves in outer space. The science of radio astronomy began to be used to probe the universe in ways that light-telescopes could not, and, as a result, various discoveries were made ... including the existence of Cosmic Microwave Background Radiation.

During the year that Alfvén received his Nobel Prize (1970), the X-ray telescope Uhuru was launched. Other X-ray telescopes became operational in subsequent years, including the Chandra (July 23, 1999) and XMM-Newton (December 10, 1999).

Before the foregoing instruments were released, X-ray activity in outer space was considered by many scientists to be a fairly restricted phenomenon. However, after the foregoing packages were launched, X-ray activity appeared to be ubiquitous across the cosmos – even including regions between galaxies.

X-rays constitute charged particles that are accelerated to extremely high velocities. For the most part, the foregoing acceleration process is made possible through the activity of electromagnetic fields that course through plasmas in intergalactic space.

In addition to X-rays, there is also a set of electromagnetic emissions that are known as Synchrotron radiation. This consists of electrons being accelerated – again by means of the electromagnetic forces that arise in conjunction with plasmas – to velocities that are at near light speeds while spiraling in curved paths through magnetic fields.

Before Synchrotron radiation was detected in conjunction with astronomical observations (they previously had been detected in relation to Earth-bound particle accelerators), Alfvén had predicted the existence of such forms of cosmic radiation. That prediction was rooted in his understanding of the role electromagnetic forces played in the plasmas that permeated the universe, and that perspective was confirmed in 1956.

For years, proponents of the standard model of cosmology have resisted the idea that charge separation could occur in space. Consequently, they felt that the impact of electrical effects in the cosmos could be disregarded without appreciably – if at all -- affecting their models in any problematic way.

Nonetheless, over the years, increasingly complex and sophisticated instrumentation packages were being launched. Many of these packages were generating data that indicated the universe was alive with the presence of plasma-based electromagnetic activity.

Consider the question: What holds galaxies together? Do gravitational forces serve to hold things together, or do electromagnetic forces constitute the predominant binding force that holds together the indefinitely large number of stars that are found in galaxies, clusters and superclusters?

The existence of millions and billions of stars in a galaxy or cluster or supercluster might seem to make such a cosmic structure a rather closely-knit and intimate setting where all those stars tug on one another and, thereby, help to keep things relatively compact. Nonetheless, the distance between stars – even closely situated ones – involves a number of light years, and, therefore – like an atom -- there is a considerable degree of cosmic structure that involves just space.

One could take an object the size of the sun (880,000 miles in diameter) and catapult it toward any galaxy one desires (assuming, of course, one had the means to accomplish this), and the likelihood that the sun would collide with any stellar member of that galaxy is extremely small (approximately in the order of 1 in 4 x $10^8$ chances). That probability is predominantly shaped by the light years of emptiness that exist in any given galaxy, cluster, or supercluster.

Such space is not empty. It is often filled with plasmas that have the capacity to generate extensive and extremely powerful electromagnetic activity.

As is true in relation to the force of gravity, electromagnetic activity exhibits a capacity to attract objects, but unlike gravity, electromagnetic activity also has the capacity to repel objects. Furthermore, as indicated previously, the strength of electromagnetism is considered by some to be 39 orders of magnitude greater than the strength of gravity (there are also some scientists who question whether electromagnetism is 39 orders of magnitude stronger than gravity, or whether, given the right scale – e.g., the Planck mass -- gravity might be 137 times stronger than electromagnetism is).

In addition, one might keep in mind that the force of gravity falls off with the square of the distance, while the force of a magnetic field that has been created by an electric current diminishes inversely as the first power of the distance from the current. As a result, the electromagnetic activity that is taking place in relation to cosmic plasmas is capable of maintaining a presence that lasts longer than (or does not drop off as quickly as) the force of gravity does.

Relative to gravity, the currents of electromagnetism are more complex (i.e., can both attract and repel), last longer, and are considerably stronger as far as their potential for organizing galactic structures is concerned. Indeed, one might suppose that it appears to make greater sense to maintain that the stars in a galaxy are more likely to be responsive to the electromagnetically active plasmas that surround them than such stars are likely to be responsive to sources of gravity that -- even if considerable -- are, nonetheless, many light years distant and exhibit a strength that falls off with the square of the distance.

Back in 1933, the astronomer, Fred Zwicky, was conducting research involving the calculation of velocities for galaxies that were part of what was known as the Coma Cluster. He calculated that the galaxy velocities were sufficiently high that the cluster should have broken apart, but this was not the case.

In order to try to explain why things weren't breaking apart, Zwicky performed some calculations that estimated the total mass of the galaxies. However, the figure he came up with for that overall mass was far too small to generate a sufficient level of centripetal force that would have been capable of keeping the Coma cluster intact.

Faced with a mystery, Zwicky hypothesized that maybe there was a form of mass present in the Coma cluster galaxies that was hard to detect because, for whatever reason, it didn't give off enough light to be seen by astronomers. He referred to such mass as "dark matter".

Using the tools of Newtonian physics, one can calculate how much missing mass would be needed to produce the Coma cluster phenomenon. Moreover, one can apply such techniques to the universe as a whole – at least the parts that are accessible to us – and theorists have come up with calculations indicating there might be as much as 23% of the universe that exists in the form of the foregoing sort of missing mass.

Consider another measurement that leads to problematic results. 'Rotation Curves" plot a star's tangential velocity in relation to its distance from the core of a spiral galaxy.

Generally speaking, the greater the distance that an object is from a source of gravity (such as in the case of our solar system), the more slowly such distant objects will travel in their orbit around the center of gravity. However, the behavior of stars in galaxies does not seem to conform to the foregoing principle of Newtonian physics.

With the exception of stars that are located near the central portion of a galaxy, most of the stars in a galaxy seem to travel with the same velocity ... even those stars that are located nearer to the edge of those galaxies. In order for a star's tangential velocity to remain constant, the amount of mass – collectively considered – must increase proportionally to the radius, but as one approaches the 'halo' regions of the galaxy (that is, the outer edges), the number of stars decrease, and, therefore, mass does not increase proportionally to the radius.

One appears to be left with a Hobson's choice. One can either assume that, for some unknown reason, mass is missing from the halo regions of a galaxy, or one can assume that there is something wrong with both Newton's theory of gravitation and Einstein's theory of General Relativity, and neither option seems all that inviting.

There is alternative approach that might be considered as a way to account for why the tangential velocity of many stars in the regions beyond the center of a galaxy appears to remain fairly constant even when located near the edge of the galaxy ... a behavior that appears to run contrary to the requirements of Newtonian physics. Perhaps the electromagnetically charged plasmas that permeate a galaxy operate like a medium that -- with the exception of stars located near the center of the galaxy -- tends to regulate the speed in a constant fashion as stars move through that kind of medium.

Star masses might decrease in numbers as one moves toward the edges of a galaxy. However, plasma activity doesn't necessarily decrease in the same way that star mass does as one moves toward the galactic halo regions.

If electromagnetically charged plasma activity were responsible for regulating the velocity of stars as they move about a galaxy, then one is not necessarily forced to choose between two problematic alternatives that either require one to explain the issue of missing mass (e.g., by inventing the existence of "dark matter) or require one to explain why the physics of Newton and Einstein appear to break down under such circumstances. The amount of energy, as well as attractive and repulsive forces, that can be stored, released, and channeled by plasmas is enormous, but astronomers and cosmologists have been reluctant to explore such possibilities.

As long as cosmological theorists insist that only the forces of gravity must be called upon to account for what takes place on the cosmic scale of the universe, then, such theorists might be working at a disadvantage. This is especially the case given that there is an ever-increasing amount of scientific evidence indicating how prevalent plasmas and concomitant electromagnetic activity are in the universe.

Dark matter is a hypothetical concept. No one has, yet, discovered the existence of one atom or particle of dark matter.

Plasmas are not hypothetical in character. The electromagnetic activity that takes place in conjunction with plasmas is also not hypothetical in character.

What needs to be determined is whether such realities have the potential – either on their own or in combination with gravitational forces – to account for the aforementioned anomalous data associated with "Rotation Curves" of stars within galaxies or with the behavior of galaxies in clusters. Before inventing mysterious forms of matter or jettisoning the physics of Newton, perhaps cosmologists and astronomers should take a closer look at something that actively permeates the universe.

One might note in passing that many of the phenomena associated with one of the darlings of modern astrophysics -- namely, black holes – might have more to do with electromagnetism than gravitation. For example, certain galaxies and stars have been observed that are producing intense jets of energy, accompanied by synchrotron radiation, and powerful magnetic fields.

The foregoing phenomena are attributed to the activity of unseen black holes that might be located in the galaxies or near stars that are displaying those kinds of energy jets and synchrotron radiation. However, no one knows – although their are any number of theories about -- how black holes would give rise to such jets, synchrotron energy, and magnetic fields.

Black holes might, or might not exist. To date, no one has seen a black hole, but an array of theoretical considerations and astrophysical data are being interpreted as being consistent with the existence of such mysterious "holes".

However, even if black holes do exist, the aforementioned phenomena of intense energy jets, synchrotron radiation, and powerful magnetic fields might be a function of something other than, or in addition to, black holes. For example, all of the foregoing phenomena can arise in conjunction with plasma activity, and, therefore, at the very least, one might have to consider the possibility that black holes – if they exist -- interact with plasmas in a variety of ways and, in the process, give rise to some of the extremely powerful – but not currently understood -- processes that have been observed in the cosmos and, heretofore, have been attributed to the mysterious activities of black holes (More will be said in conjunction with the topic of black holes in a later chapter of this book).

According to the mainstream model of cosmology, black holes are theorized to be the end result of an evolutionary trajectory for stars that satisfy a certain set of conditions. Let's take a look at the way in which the current standard model of cosmology describes the nature of stars in general.

Today, many astrophysicists believe that over the course of millions of years in the early universe, gravitational forces eventually were able to draw together sufficient quantities of (mostly) hydrogen gas and various particles to enable such nebulae to pass through a critical threshold that ignited a self-sustaining process of nucleosynthesis in which atoms that are more complex than hydrogen began to be generated. According to mainstream astrophysics, the core of a fledgling star is a nuclear reactor that fuses hydrogen atoms together in a way that leads to the emergence of helium atoms, along with the release of a tremendous amount of heat that is radiated away from the core by photons that undertake a journey of several hundred thousand years, or so, toward the surface of the star.

The core of a star occupies approximately 20% of a star's radius. That core is encompassed by a region, believed to occupy another 50% of the star's radius, which consists largely of photons radiating away from the star's core (but with no movement of materials), and the remaining 30% of the star is considered to serve as a convection zone (involving the movement of materials) that transports heat energy to the surface of the star where photons are released from a long journey that began at the core.

The outer, surface area of the aforementioned convection zone is known as the photosphere. Extending out from the photosphere is a region that is between 1250 to 1900 miles and is referred to as the chromosphere.

The corona of a star forms along the outer boundary of the chromosphere. Beyond the corona is an extensive region reaching far out into the surrounding space for billions of miles, and that zone is the plasmasphere.

If the standard model of a star's mode of behavior were correct, then, the activity of the chromosphere, corona, and plasmasphere should give expression to a fairly simple scenario in relation to what allegedly transpires in those regions. More specifically, there should be nothing going on in those zones other than the radiation of heat and light away from the star.

This is not the case. The processes occurring in the aforementioned regions are far more complex than the standard model of a star's life predicts is the case.

For example, the lower portion of a star's corona is millions of degrees hotter than the photosphere or surface of that star. If a star's surface is merely the place were heat and light begin to radiate away from a star, then, why should the corona be so much hotter than the surface of that star?

The temperature of the Sun's visible surface – or photosphere – has been measured to be around 5800 degrees Kelvin. The temperature near the deepest part of a sunspot is even lower.

Initially, as one begins to move away from the surface of the Sun, the temperature falls. These falling temperatures have been measured to be between 3800 and 4000 degrees Kelvin.

However, in the regions near the lower coronal areas that are above where the foregoing minimal temperatures are found, the temperature picture changes dramatically. A climb toward several million degrees Kelvin begins to take place, and this constitutes a mystery for the standard model of stellar behavior.

Or, consider what might be referred to as the stellar or solar wind that sweeps away from the surface of stars. The stellar wind is actually an electric circuit that is generated by ions, and, within limits, the velocity of those ions actually increases the further away from the star they get.

Radiant energy supposedly operates in accordance with an inverse square law which indicates that the further one recedes from a heat source, then, the less energy should be present and the lower the temperature should be. Yet, a corona that is thousands of miles from the surface of a star has a temperature that is millions of degrees hotter than its surface, and, in addition, ions in a so-called 'stellar wind' actually accelerate faster the further they are from a star's surface.

The radiation model that is inherent in the standard cosmological model of a star's behavior cannot account for either of the foregoing phenomena. However, plasmas that generate electromagnetic activity are able to provide a plausible account of those observations.

For instance, due to the dynamics of charged particles in plasmas "drift currents" – or voltage gradients – form. As ions become caught up in the drift currents that lead away from a star, the ions begin to accelerate ... something that is observed to occur in relation to so-called solar winds.

Those who are less enamored with the possibilities of plasma physics suggest an alternative account for the foregoing phenomenon. For example, they claim that the photons being released from a star's surface collide with ions and electrons, and in the process, accelerate the latter.

However, such an approach doesn't necessarily explain why the 'solar wind' particles accelerate faster as they recede from the surface of the sun. This behavior seems more consistent with the drift current notion of the voltage gradients created in plasmas than it appears to be consistent with the inverse square law that, as some claim, governs the energy of photons as the latter particles move away from the sun.

Turning, now, to the issue of temperature differentials between the photosphere and the lower portion of the corona, from the perspective of electromagnetic plasmas, positive ions become first 'dethemalized' and, as a result, are less chaotic or random in their movements ... and this translates into lower temperatures. However, as the positive ions become caught up in the aforementioned voltage gradients that have been generated in the plasma, the positive ions begin to speed up and, consequently, their motion becomes more violently chaotic as they interact with other equally energetic positive ions, leading to higher and higher temperatures as one moves through the voltage gradient.

Other facets of a star's behavior also have been filtered through the lenses of the electromagnetic activity that occurs in plasmas. For example, in the 1970s, Ralph Juergens theorized that a star's power might not be a function of the fusion reaction processes that allegedly are taking place in the core of a star. More specifically, according to Juergens, stars possess a significant electric capacitance (that is, the capacity to hold an electric charge).

Juergens contends that the galactic currents that are running through the plasma surrounding a star (the plasmasphere) feed the aforementioned capacitance. More specifically, the Sun acquires a positive electrical potential relative to the plasma surrounding the Sun.

As a result, positive ions are transported away from the Sun while negative electrons flow to the Sun. This generates an electrical current that flows through the poles and exits from lower latitudes of the Sun.

Juergens performed calculations in conjunction with some estimates of both the Sun's voltage and the electrical currents that might be flowing to the Sun through the plasmasphere ... electric currents that are coming from the surrounding galaxy. His calculations led him to hypothesize that the Sun was receiving sufficient energy from galactic currents and was producing sufficient voltage to give expression to a power output that is comparable to what is observed in relation to the Sun, and, therefore, Juergens maintained that the Sun (and, consequently, any star) might be powered by galactic plasma currents surrounding a star rather than from some sort of internal nuclear fusion process deep in that star.

Juergens doesn't believe that processes of nuclear fusion are occurring in the core of a star. He contends that the Sun – and, therefore, other stars – is (are) isodense in character ... that is, having approximately the same density throughout.

Yet, if stars are isodense and there are no fusion processes taking place in the core of a star, then, how does nucleosynthesis take place? Some plasma physicists believe that Birkeland currents operating in arc mode are sufficiently strong to produce z-pinch effects among the filaments of the Sun's photosphere and, in the process, fuse atoms.

The foregoing z-pinch effects would be especially prominent in the double layer regions near the top of the photosphere. Such double layers constitute regions where electromagnetic forces in plasmas are often the most powerful.

Scores of elements (more than 60 of the 92 naturally occurring elements) have been observed in the spectroscopic analysis of the Sun's light. If plasma physicists are correct, those elements could have been fused in regions near to the photosphere rather than fused deep in the core of the Sun.

Mainstream astrophysicists claim that traces of the elements that are found in the spectroscopic analysis of the sun's light constitute elements that have been generated through supernovae that have dusted the Milky Way – including the Sun -- with an array of elements over the course of millions of years. Yet, even if the foregoing aspect of the mainstream account is accurate, that explanation still leaves unanswered the issue of how such elements are generated in the first place.

Have those elements been produced within the core of various stars (cores that have not been probed empirically to determine what is actually taking place there)? Or, have those elements been synthesized somewhere in the vicinity of the Sun's (or a star's) photosphere?

The foregoing possibilities constitute competing theories concerning the process of nucleosynthesis in stars. Currently, most astrophysicists prefer the core-approach rather than the surface-approach to nucleosynthesis, but the issue is far from being definitively settled.

Let's take a brief look at another set of competing arguments involving mainstream astrophysics and plasma physics. For example, if – as Juergens claims -- there are no processes of nuclear fusion that are taking place in the core of a star, why doesn't the star experience a gravitational collapse?

According to the perspective of mainstream astrophysics, there is an outward push of photon radiation that is generated through the process of nuclear fusion at a star's core. This outward pressure of radiation serves as a countervailing force in relation to the gravitational attractive pressures that exist within a star.

Plasma physicists, on the other hand, contend that a star consists of a set of plasma currents. Electric dipoles are created within those plasmas and align with one another to form radial electric fields that induce electrons, in great quantities, to move toward the surface.

The forgoing exodus of electrons results in a predominance of positive ions in regions previously containing electrons that had been in an electric dipole arrangement with positive ions. The like-charged ions that have been left behind repel one another, and the force of such repulsion (which is 39 -- ?? -- magnitudes stronger than gravity) is enough to offset the presence of the gravitational forces of attraction that exist within a star.

How does one decide between two competing theories involving an array of issues such as have been touched upon in the foregoing pages? Presumably, the best way to settle matters is by means of empirical data.

For instance, critics of Juergens' general model of the Sun (and, therefore, model of stars in general) argue that the kind of energy flows toward the Sun that is required by his theory have not been observed. However, extended Birkeland currents were discovered through the instrumentation on board Ulysses ... a vehicle that has now been decommissioned but had been launched in 1990 to study the Sun's activities at various latitudes.

The Ulysses project made three latitude-scans of the Sun. These scans took place in 1994/1995, 2000/0001, and 2007/2008.

The scan of most relevance to the present discussion occurred in 2000/2001. It was during the latter scan that the aforementioned Birkeland currents or plasma tubes were observed, extending from the south pole of the Sun to the vicinity of Mars.

Could such 'plasma tubes' be a means of funneling energy to the Sun in accordance with Juergens model? Is it possible, as well, that similar sorts of tubes might link the north pole of the Sun with plasma currents that are generated as the Sun moves about its orbit within the Milky Way?

The foregoing questions have not been resolved in any final sense. However, until definitive answers are determined, one cannot automatically ignore ideas like those of Juergens just because they constitute a challenge to what is considered to be the mainstream view of astrophysics.

Another source of data that might shed light on whether mainstream or plasma accounts of astrophysics give expression to better or more consistent models involves sunspots. The umbra (the dark, center portion) of a sunspot is cooler by approximately several thousand degrees Kelvin relative to the surrounding photosphere (3800 to 4000 degrees Kelvin and 5800 degrees Kelvin respectively).

Conceivably, sunspots are holes in the photosphere. If so, then, sunspots would provide a window of sorts into some of what might be occurring just beneath the photosphere

For example, if – as mainstream theories hold – energy flows from the core of a star (with temperatures in the vicinity of fifteen million degrees Kelvin) to a star's surface by means of, first, radiation zones, and, then, convection zones, one might suppose that sunspots represent the top portion of the convection zones that transition into the photosphere or surface potions of a star. Furthermore, if sunspots do constitute the top of the aforementioned convection zones, then, one would expect them to be hotter than the photosphere, not cooler.

Mainstream theories often allude to various factors – such as magnetic fields – occurring beneath the photosphere to account for the temperature differential between the umbra of a sunspot and the surrounding photosphere. Such factors are believed to interfere with the release of energy and heat from a star's interior, thus, resulting in sunspots.

While the foregoing possibilities might be correct, offering an explanation is not the same thing as verifying what is being claimed. Perhaps, the reason why sunspots are cooler than the surrounding photosphere is due to the way various kinds of magnetic fields interfere with the journey of heat and energy to the surface of a star, but, it is also possible that the reason why sunspots are cooler than the surrounding photosphere is because the interior of stars (especially the core) are not as hot as mainstream theories believe is the case.

The temperature of a star's interior is set by theory ... not observation. For example, the temperature for the core of a star is a function -- in part -- of what is needed to establish a self-sustaining process of nuclear fusion through which nucleosynthesis would be able to take place.

If there is no process of nuclear fusion occurring in the core of a star – as some theorists such as Juergens hypothesize – then, there is no need to estimate that the temperature of the core interior of a star must be at least as high as would be needed to underwrite a process of self-sustaining nuclear fusion. Moreover, if the interior of a star is not as hot as mainstream theory claims, then, it makes a certain amount of sense to argue that sunspots are several thousand degrees Kelvin cooler than the photosphere because the latter is an expression of a surface plasma physics that is different (and hotter) than the sort of plasma activity that is occurring in the interior of a star ... especially if the aforementioned surface physics is fed through Birkeland currents that are being funneled to a star's surface due to the interactional nature of the activity occurring between the plasma fields surrounding a star and the star itself.

In addition, some plasma physicists contend that sunspots encompass regions in which the usual activities of the photosphere are no longer taking place and, as a result, a normal barrier to the upward flow of positive ions has been removed. Consequently, in sunspot regions, the upward flow of positive ions toward the zone of the corona establishes a substantial electric current that, in turn will generate a magnetic field in relation to the sunspot.

Due to the way in which the energetic positive ions that are streaming up through the sunspot collide with the atoms present in the lower corona region, powerful X-rays are produced. Some of the brightest portions of X-ray images – portions that indicate hot, energetic activity – are located above high concentrations of sunspot activity, and, therefore, such images are consistent with what might be predicted by a plasma model of a star's behavior.

Consider another source of data. In 1995, a joint project involving NASA and the European Space Agency led to the launching of SOHO (Solar and Heliospheric Observatory), and about six months later, the project became operational.

Despite running into some problems that almost led to the demise of the project, SOHO is still up and running. In fact, its mission has been extended until at least some time in 2016.

There were a dozen, or so, instrument packages on board the observatory. Those instruments were geared to study an array of phenomena, ranging from: Solar winds, to: Coronal physics and solar core oscillations.

Among the many observations that have emerged from the research conducted in conjunction with SOHO is the following. There are rivers of plasma running beneath the photosphere of the Sun.

These plasmas are electrically charged. Moving, electrically charged plasmas generate magnetic fields.

Some of the magnetic fields that are produced form omega-shaped loops that extend outward into the chromosphere (above the photosphere). Visually, one cannot see the magnetic fields, but one can observe the glowing Birkeland currents or filaments that form along, and follow, those magnetic fields.

Under certain circumstances (e.g., if the associated voltage gradient becomes sufficiently strong), the circuits that exist become disrupted or broken. When this occurs, the energy stored in the magnetic field is explosively released in the form of flares and mass coronal ejections.

A considerable time before SOHO was launched, Hannes Alfvén had developed a model of electric circuitry in which currents runs through each of the poles of a star. This leads, in turn, to the formation of secondary currents that run along, or just beneath, a star's surface and toward the equator.

The foregoing circuitry induces magnetic fields. These give rise to bipolar magnetic regions that bulge outward through the photosphere of a star.

The foregoing magnetic regions can store and release energy under shifting conditions of electromagnetic activity involving flowing rivers of plasma. All of this is consistent with what SOHO brought to light.

In 1999, Researchers involved with the previously mentioned Ulysses space project indicated that the magnetic field of the Sun is becoming increasingly stronger. In fact, available evidence from a variety of sources indicates that the strength of that field has doubled over the last 100 years, or so.

Treating the Sun (or stars in general) as being largely fusion reactors doesn't account for any of the foregoing data. While future evidence might conclusively demonstrate that nucleosynthesis takes place in the cores of stars (and, remember that the idea of core-based nucleosynthesis is a theoretical account that might, or might not, be correct), nonetheless, even if such core-based nucleosynthesis is occurring, this, clearly, might be only part of the story as far as what transpires in stars is concerned.

Conceptualizing stars as being, for the most part, nuclear fusion reactors tends to ignore the theoretical possibilities (many of which are consistent with collected data) that are entailed by plasma physics. Indeed, plasma physics appears to give expression to plausible modes of interpreting many of the behaviors that have been observed in the Sun and, therefore, presumably, will be observed in other stars as well.

The present chapter has barely scratched the surface concerning what plasma physics might have to offer to the discipline of cosmology. That potential seems considerable, and, yet, unfortunately, many prospective cosmologists and astrophysicists apparently receive very little exposure to such ideas during their formative, educational years.

From the perspective of the 'Final Jeopardy' challenge and the concomitant reality problem that forms the woof and warp shaping the series of volumes in which the present book plays a part, denying individuals (whether done intentionally or unintentionally) the opportunity to explore ideas, methods, and understandings that appear to have relevance to the process of helping to push back the horizons of ignorance concerning the nature of reality seems inexcusable. All too much of what takes place with respect to cosmological research and education appears to be excessively entangled in the strings of bias, conformity, and vested interests than such research and education seems to be really dedicated to a sincere, open, and rigorous search for truths concerning the nature of reality … and, as the first three volumes of *Final Jeopardy* have documented fairly extensively, the foregoing difficulties also occur in many other scientific disciplines as well.

Without a commitment to truth, we begin at no beginning, and we work toward no end. Without a commitment to truth, any answer that one might give in conjunction with the Final Jeopardy challenge is likely to be sloppy, error-ridden, and problematic.

Plasma physics might not be the answer to all problems. Nonetheless, that discipline appears to constitute a means of reasonably and, sometimes, successfully addressing quite a few more problems in astrophysics concerning the nature of reality than mainstream cosmology appears to be inclined to acknowledge at the present time.

## Chapter 5: Matters of Gravity

Ancient, human eyes looked up into the night skies and saw the mysterious occupants of space give expression to a combination of regular and irregular behaviors. On the one hand, there were five "stars" (i.e., planets) that appeared to wander from place to place in complicated paths, and, on the other hand, there were the "fixed" stars that could be relied on to behave in predictable ways during the course of a year as well over the course of a lifetime.

The foregoing regularities and irregularities were eventually organized into a geocentric theory. The motion of heavenly bodies became somewhat predictable and understandable if one were to consider that behavior as the orbiting of bodies about a stationary Earth.

The chief architect of the geocentric model was the Roman, Ptolemy. He lived in Alexandria, Egypt during the 2$^{nd}$ century.

Although his theory eventually was overturned, the change in human understanding concerning the physical nature of the universe took some 1400 years to arrive. One reason why the revolution in astronomy was so long in coming was because, in many respects, the Ptolemaic system worked.

In other words, serviceable predictions could be made using the Ptolemaic model. Moreover, if problems arose, many of those difficulties could be resolved by adding the right number and kind of epicycles (small circles whose center moved along the circumference of larger circles) to account for the motion of cosmic bodies.

Another reason why the Ptolemaic system survived for so long had to do with its rootedness in geometric figures that had an almost hypnotic hold on the human mind – namely, the seeming perfection of spheres and circles. This conceptually aesthetic attraction helped to keep alive the great mathematical/philosophical traditions of Pythagoras and Euclid.

Nearly 400 years prior to Ptolemy, someone devised a very different model of the universe. More specifically, on the basis of a variety of observations that had been made over time by different people, and by employing established geometrical techniques, Aristarchus – who lived in Samos, Greece -- not only was able to calculate the sizes of the Sun, the Earth, and the Moon, but, as well, he developed a heliocentric model in which the Earth revolved about the Sun.

Perhaps due, at least in part, to the eclipse of Greece's prominence by the rise of the Roman Empire, along with the increasing influence of certain dimensions of Christian theology, the heliocentric ideas of Aristarchus disappeared. The geocentric model of Ptolemy came to dominate the thinking of much of the so-called civilized world.

Prior to the paradigm-shift that started in 1543 with the appearance of the book: *De Revolutionibus Orbium Coelestium* (*On the Revolutions of the Celestial Spheres*) by Nicolaus Copernicus, there had been several individuals who toyed, in rather tentative ways, with the idea of a heliocentric model concerning the relationship between the Earth and the Sun. For example, in 499 A.D. Aryabhata, an Indian astronomer, developed ideas involving an Earth that spun on its axis and, as well, advanced a model in which planetary periods were considered in relation to the position of the Sun relative to those periods.

In addition, two Arab thinkers – Abu al-Rayhan al-Biruni, around 1000 A.D. and Najm al-Din al-Qazwini al-Katibi, some 300 years later – both entertained ideas that were similar to those of Aryabhata before them. However, each of the aforementioned Arab thinkers abandoned that kind of model … perhaps daunted by the pervasive intellectual pressure that flowed through the dominant model of the day: the geocentric model of Ptolemy.

There is evidence to indicate that Copernicus was influenced – at least partially – by his predecessor al-Katibi since Copernicus used diagrams that were labeled in the same manner as some diagrams that were used by al-Katibi. However, Copernicus put forth more rigorous and extended arguments than his predecessor had concerning why a heliocentric model should be preferred to a geocentric system.

Subsequently, improvements were made to the heliocentric model, but, perhaps, one of the biggest improvements came when ellipses replaced the use of circles in relation to the configuration of astronomical models. As previously indicated, many people were mesmerized by what they considered to be the perfect nature of circles and spheres, and, therefore, were reluctant to jettison such ideas from their thinking.

In large part, the critical changeover in mathematical thinking concerning the issue of geometric shapes was made possible by the work of Johannes Kepler, a German mathematician and astronomer, who got his start by assisting Tycho Brahe, a Dane, who was working on planetary motions in Prague, Czechoslovakia. Brahe had collected a great deal of data on the motions of planets, and after Brahe died in 1601, Kepler made good use of that data by developing his three laws of planetary motion that were published in 1609.

Kepler's first law stipulated that all planetary orbits are elliptical in nature. His second law specified that planets plow through equal areas in equal times as they move along their elliptical orbits, and his third law of planetary motion indicates that the squares of the orbital periods of planets are proportional to the cubes of their distances from the Sun.

Although Kepler is often cited as one of the trailblazers of modern science, the fact of the matter is that the issue is not straightforward. Like many other mathematicians of his day, much of Kepler's mathematics and science were deeply embedded in head-scratching speculations involving philosophy, religion, and astrology.

Isaac Newton – who, like Kepler, was keenly interested in astrology, religion, and the occult – derived his theory of gravity from Kepler's laws of planetary motion. Unlike Kepler, Newton managed – for the most part -- to clearly separate his physics from his other non-mathematical and non-scientific interests.

At the heart of Newton's view of the universe is the law of gravity. This law states that all bodies in the universe attract one another with a force that is proportional to the product of their respective masses while, simultaneously, that force is inversely proportional to the square of the distance between such bodies. [i.e., $F = G(m_1 m_2 / d^2)$, where G constitutes the gravitational constant, $m_1$ and $m_2$ are the masses of the two bodies that are gravitationally attracted to one another, d is the distance between the bodies, and F is the gravitational force between the bodies].

Newton's law of gravity is a universal law of attraction. This property of universality has an upside and a downside.

The upside of universality is that it provides a framework through which to analyze how one force – namely, gravity – can affect all objects (both heavenly and Earthly) in the same way. The downside to universality is one can never demonstrate that all objects in the universe actually have a mutual attraction for one another.

One can, of course, take any two bodies and work out the specifics of the gravitational force between them. If the two bodies interact with one another in accordance with the law of gravity, then, one will have demonstrated that Newton's theory of gravity applies to those two bodies, and, as a result, one will have generated data that confirms Newton's model.

Nonetheless, if one were to take two bodies on opposite sides of the universe – bodies that are separated by billions and billions of light years -- the situation becomes empirically problematic. In other words, although one could use Newton's formula to calculate the theoretical force with which two bodies supposedly attract one another, one has no way of knowing whether, or not, such a mutually attractive force is present ... that is, one has no way of being able to measure whether each of the bodies actually tugs on the other with the sort of attractive force that Newton claims is universally operational.

Newton believed the universality of gravitational attraction was possible because it operated by means of instantaneous action at a distance. Newton did not know what made such "action at a distance" possible. Moreover, Newton could not verify that such action was, indeed, instantaneous.

However, in most instances, neither of the foregoing issues mattered much. Practically speaking, all one needed to do was to make the requisite calculations involving masses, distance, and the gravitational constant, for any two bodies and determine whether, or not, observation indicated that those calculations reflected Newton's equations.

Newton developed a way to describe the behavioral properties of gravitational dynamics. Nevertheless, Newton didn't actually understand the nature of gravity.

He didn't know how gravity generated its "action at a distance". Furthermore, Newton didn't know how the force of gravity was transmitted instantaneously.

Furthermore, there was a sense in which the gravitational constant that appears in Newton's law of gravity was something of a fudge factor, just as 234 years later on, Planck's calculation of the value for a quantum of energy played a similar role. In both cases, the values for the two, respective constants were arrived at because those values enabled their originators to solve problems correctly and not because either of those individuals actually understood the nature of the phenomena they were so ingeniously describing.

The gravitational constant was not actually measured in a laboratory until 71 years after Newton passed away when Henry Cavendish performed an experimental measurement in 1798 using a torsion balance devised by John Michell. However, the actual value of the gravitational constant was implicit in the Cavendish measurement rather than overtly given by that process.

The value for the gravitational constant is quite small. It is: $6.67384 \times 10^{-11}$ m$^3$ kg$^{-1}$ s$^{-2}$.

Depending on one's point of reference, gravity can be considered to be either weaker or stronger than the electromagnetic force. On the one hand, if an individual were considering the relationship between an electron and a proton, then, electromagnetism is 39 orders of magnitude stronger than the strength of gravity with respect to the nature of the interaction between the electron and the proton, but if, on the other hand, an individual engages the comparative strength of force issue through the lenses of, say, the Planck mass (approximately 22 micrograms), then there are calculations that can be performed which indicate that gravity is 137 times stronger than electromagnetism.

There might not be any way to determine – in a precise, absolute sense – the strengths of gravity and electromagnetism relative to one another. This is because: The nature of the circumstances being considered, the units of measurement being used, and the somewhat arbitrary nature of comparing dimensionless values ... such factors can all affect the process of considering the relative strengths of the two kinds of forces. However, on the quantum level, the effects of gravity tend to be negligible, and this reference point will be important later in the chapter when the idea of quantum gravity is explored.

Whatever the nature of the relative strengths of gravity and electromagnetism might be, the value of the gravitational constant entailed by the Cavendish measurement was remarkably accurate, and, therefore, only relatively slight improvements have been made to the precision of the gravitational constant over the last several hundred years. Yet, at the present time – and despite the precision of the value that has been determined for the gravitational constant -- no one understands why that constant has the value it does nor does anyone currently understand how something with that value is transmitted or exchanged between bodies.

The mysteries that permeated Newton's theory of gravity did not end with: Action at a distance, the idea of instantaneous attraction, or the nature of the source of the gravitational constant. While Newton's theory worked very well for two bodies, it became bogged down when trying to figure out solutions involving three or more bodies.

Over the years, mathematicians and scientists developed various techniques involving shortcuts, work-a-rounds, and tricks that enabled approximately correct solutions to be calculated with respect to various problems involving three bodies. Later on, computers that employed appropriate software programs were able to generate even better results – both more quickly and more easily.

Nonetheless, three body problems – whether done by hand or with a computer – often ran into difficulties that were inherent in the way in which mathematics went about engaging reality. During the latter part of the 1800s Henry Poincaré analyzed the three-body problem and uncovered the first signs of chaotic systems in which the interaction of as few as three bodies could be so complicated and entangled that their dynamics appeared to defy resolution even while those motions were completely determinate in nature.

For example, the bodies that comprise the Solar System gravitationally interact with one another in determinate ways, and one knows this because, among other things, we observe predictable regularities in the orbital motions of the bodies that exist within such a system. Yet, trying to describe the complexities of the foregoing sorts of interactions through the mathematical ideas that currently are available to us often becomes permeated with a nightmarish-like quality that escapes the capacity of such methods to definitively resolve.

While the three-body problem touches on various difficulties – some of which have been resolved and others which have not – that problem does not constitute the same kind of issue that is involved in trying to determine the nature of gravity itself. The latter problem tends to be of an essential nature, while the three-body issue is somewhat peripheral and derivative in character.

Newton was not able to penetrate beyond the surface behavior of gravity. He could describe the behavior of systems that were under the influence of gravitational forces, but he could not say what the precise nature of the force was that underwrote such gravitational behavior.

Somewhat surprisingly (at least, perhaps, for some), Einstein didn't fare much better when it came to actually understanding the nature of gravity. General relativity (Einstein's theory of gravity) did entail a more powerful and precise set of tools for describing aspects of gravitational behavior that could not be properly captured or addressed by Newton's method of analysis [e.g., the nature of the anomalous properties in the precession (the change in the orientation of the rotational axis of a rotating body) of Mercury's perihelion (the orbital point that is nearest to the Sun].

However, Einstein considered gravity to be a matter of geometry without ever explaining how that geometry came to be curved through the presence of gravity. Like Newton – although Einstein did so in his own inimical fashion – Einstein never actually discussed the actual nature of gravity but spoke only in terms of its behavioral effects.

By operationalizing gravity in terms of geometry, Einstein was able to get rid of Newton's notion of spooky action at a distance, and, consequently, the behavior of gravity became a function of the properties of local geometry. Furthermore, Einstein eliminated Newton's idea of instantaneous gravitational influence by limiting the rate of gravitational transmission to the speed of light, and, thereby, reconciling gravity with the special theory of relativity.

Nevertheless, Einstein did not understand the inner character of gravity any better than Newton did. For both of these individuals, the actual nature of gravity was a complete mystery as far as what made gravity possible or what was being transmitted through gravitational attraction or how such transmission actually took place.

In a sense, Einstein's fruitless search for a unified theory of physics during the last several decades of his life appears to be – at least to a degree -- an acknowledgement that there were some significant lacunae present in his understanding of how nature worked. His attempt at unification was an effort to incorporate electromagnetism and gravity into one mathematical framework, and this effort, if successful, might shed light on, among other things, how gravity did what it did ... something that his general theory of relativity was not able to accomplish.

Of course, Einstein was operating under a bit of a handicap. The strong and weak forces – not to mention the Higgs field -- had not, yet, come into focus within the world of physics, and, as a result, Einstein was unaware that there were forces present in nature that went beyond the realms of electromagnetism and gravity and, therefore, were capable of thwarting Einstein's attempt to devise a unified theory of physics.

Since Einstein passed from the scene, some progress has been made toward unifying physics. However, at the present time, the state of the unification issue is not all that much better than was the case in 1955 at the end of Einstein's life.

Links between electromagnetism and the weak force have been established through the framework of an electroweak theory that was forged in the 1970s and 1980s. Nonetheless, the unifying links – if any – among the nuclear strong force, the force of gravitation, the Higgs field, as well as electromagnetism and the weak force continue to be elusive.

One might also keep in mind that if there was no Big Bang, then the symmetry-breaking event that supposedly brought about the differentiation of the electromagnetic force and the weak force from an underlying unified force of some kind becomes rather tenuous. In other words, if the Big Bang did not occur, whatever the connections might be between electromagnetic and weak forces, then those two forces might not have entered the universe in the way that electroweak theory envisions -- that is, as symmetry-broken remnants of a previously unified force of some kind.

Conceivably, therefore, if there is an ultimate theory of everything, perhaps electromagnetism and the weak force are not necessarily ultimately connected in the way electroweak theory suggests. As such, the electroweak theory might be – to some degree – a theoretical artifact of a set of dynamics (the Big Bang scenario) that never actually occurred.

In fact, *if* there were no symmetry-breaking event in the early universe that occurred in concert with, or because of, a Big Bang of some kind, then, the electroweak theory entails something of a mystery. More specifically, one wonders why electromagnetic forces and weak forces would have had the capacity to be unified under conditions that might never actually have existed.

Alternatively, there might have been some kind of symmetry-breaking event involving electroweak theory that did occur but was unrelated to conditions involving a Big Bang. However, if the Big Bang did not happen, and, therefore, if the nature of the foregoing symmetry-breaking event was not necessarily tied to the energies and temperatures associated with a Big Bang scenario, then, such a possibility leaves the question open as to what such a symmetry-breaking event might have involved.

Both Newton's theory of gravitation and Einstein's theory of gravitation gave expression to frameworks that focused on the effects of gravity rather than on the nature of the phenomenon that could make those effects possible. Einstein's framework had the capacity to deal with a number of gravitational effects that could not be effectively handled by Newton's system of thought, but with the exception of this enhanced dimension of gravitational theory, Newton and Einstein's frameworks produced comparable results for an array of everyday problems involving the phenomenon of gravity.

The theories of gravitation envisioned by Newton and Einstein each possessed a dimension of universality. However, the idea of universality was different in each case.

For Newton, universality was an expression of the way in which every body in the universe instantaneously and simultaneously exerted a gravitational tug on every other body in the universe. For Einstein, on the other hand, the idea of universality in general relativity paralleled the principle of universality inherent in special relativity ... namely, that the laws of physics operated in the same fashion everywhere in the universe.

Newton and Einstein also differed about their respective understandings of how – of if -- space, time, and gravitation interacted with one another. For Newton, space and time were absolute, unchangeable, and independent -- that is, not affected by one another, as well as independent of the rest of reality (and, therefore, space and time were not affected by gravity) -- whereas for Einstein, space and time were relative and changeable phenomena that could be affected by the presence of gravitational fields.

Prior to the mathematical work of Carl Friedrich Gauss, János Bolyai, Nikolai Lobachevski, and Georg Friedrich Riemann during the 1800s, space often tended to be described through the lenses of plane geometry. In plane geometry, the interior angles of triangles add up to 180 degrees, parallel lines do not meet, and the shortest path between two points is a straight line.

During the nineteenth century, a number of individuals (noted toward the beginning of the last paragraph) began to explore the properties of geometries that used different axioms than the ones with which Euclid began. In the new geometries, the interior angles of triangles could be greater or lesser than 180 degrees, parallel lines might meet (possibly more than once), and the shortest distance between two points is not necessarily a straight line in Euclid's sense of the word.

Toward the beginning of the 20th century, geometries became even more complex with the introduction of dimensions beyond the usual three of spherical geometry. For instance, until Theodor Kaluza (1919 – 1921) and Oskar Klein (1926) opened the conceptual sluice gates that were controlling discussions concerning the idea of dimensionality, one of the few individuals who thought about space and time in terms of more than three dimensions was Hermann Minkowski.

At the 80th Assembly of German Scientists and Physicians (1908), Minkowski used two lines to represent four dimensions. A vertical axis constituted time, while a horizontal axis gave collective expression to the usual three spatial dimensions. The space described by the two-axis system was a non-curving amalgamation of space and time that became known as: "Minkowski space".

A point in Minkowski space marks the intersection of time and space and is referred to as an event ... that is, such a point constitutes a marker for whatever does, or does not, occur at the intersection of time and space that is contained within the aforementioned point. Moreover, one can represent the movement of an object through space-time by drawing the 'world line' that encompasses the trajectory that is made up of a series of such point-events.

A world line could be related to the speed of light through the angle at which it was tilted in Minkowski space. Light travels with the fastest possible speed, and, therefore, a world line representing something traveling at such a speed would display a 45-degree angle in space-time,

Minkowski referred to objects traveling at the speed of light as being light-like. Objects that traveled slower than the speed of light were representing as tilting more toward the vertical – time -- axis, and were referred to as being time-like.

According to Newton, as long as no external forces are acting upon a given object, then that object will continue on in a uniform state of motion. This is the principle of inertia at work.

Uniform motion is characterized by the quality of constancy. That is, there is no change in velocity, and, therefore, there is no acceleration present in an object moving with uniform motion.

Minkowski merged Newton's principle of inertia with the constancy of motion. In doing so, Minkowski came up with the concept of a 'straight world line' ... in other words objects tend to move through space-time along a straight line trajectory in the absence of forces acting upon them.

Einstein generalized Minkowski's perspective by introducing the idea of curved space-time into the picture. If space-time is flat, then, objects operate in accordance with Minkowski's straight world line notion, but if space-time is curved (such as when in the presence of massive bodies), then objects move in accordance with the geodesics of space-time ... that is, according to the shortest world line trajectory that is possible under conditions of space-time that are curved in a given way.

Minkowski indicated that the presence of curved world lines meant that some sort of force is acting on an object that is moving along such a trajectory in space-time. According to Einstein, objects that are conforming to the geodesics of curved space-time behave exactly like objects that are moving through a gravitational field.

Consequently, gravitational force can be considered to be equivalent to the degree of curvature in space-time. Moreover, geodesics mark the path or trajectory of an object that is moving under the influence of such a force.

For Newton, gravitational force is functionally tied to mass because, for him, gravitational attraction is proportional to mass. However, Newton didn't understand the nature of the dynamic that gave rise to such a proportionality ... that is, Newton didn't understand precisely what the character of the relation is between gravity and mass other than that the former was proportional to the latter.

For Einstein, gravitational force is functionally tied to the curvature of space-time because, for him, the presence of gravity is reflected in the properties of an object's geodesic moving through such curvature. Nonetheless, although Einstein understood that gravity had something to do with mass, he didn't understand the nature of the dynamic that gave rise to the curvature of space-time in the presence of mass.

Einstein had improved on Newton's mousetrap because Einstein's framework permitted one to capture more of the subtleties and nuances of motion or behavior according to the manner in which the presence of gravity affected the way objects scurried about one or another geodesic. However, Einstein was as mystified about the nature of gravity as Newton was.

The warping or curvature of space-time marked the presence of gravity. Many people – including Einstein -- have taken the foregoing concept and generated a hermeneutical framework that is not necessarily consistent with the idea of space-time.

Space-time is a representational metric. Its only ontological dimension resides in its methodological properties.

One uses space-time to fix the position of an event (in Minkowski's sense) or to keep track of a set of events that mark the world line or trajectory of an object as it moves (event by event) through such a metric.

A metric constitutes the way a given system measures changes in a given object. Therefore, space-time provides a means of measuring how an object changes position (or not) over the passage of time, but space-time is not ontologically equivalent to space and time because the former (space-time) is a system of measurement, while the latter (space and time) are basic ontological components of physical reality.

What warps space-time does not necessarily have any impact on either space or time. Neither Einstein nor any of his acolytes have shown that curvatures in space-time translate into curvatures of either space or time.

Yes, clocks have been shown to slow down in gravitational fields, but this only means that gravitational fields have an impact on the frequency with which clocks run, and the rate at which clocks run does not demonstrate that gravitational fields actually affect time. Moreover, neither Einstein nor any of his supporters have demonstrated that the curvature of a gravitational field is the same thing as saying that space has been curved by the presence of such a field.

It is one thing to claim that the effects of a gravitational field on an object can be measured by the degree to which the curvature that is described through a space-time coordinate system affects the trajectory of an object that is described in terms of that same coordinate system. However, claiming that the curvature that is present in a gravitational field is tantamount to the curvature of space is quite another matter.

Einstein never figured out how gravity makes curvature in space-time possible. Figuring out how – or if -- gravity curves space and/or time is an even more difficult problem.

To say that gravity is capable of warping or altering space and time indicates that we know what time and space are ... that we understand, for instance, how the tapestry of space and/or time are woven from a given set of properties -- x, y, z, and so on. Furthermore, to claim that gravity is capable of warping/curving space and time also suggests that we know what gravity is and that we understand how it interacts with space and time in a way that will lead to the warping or altering of whatever set of properties (e.g., x, y, and z) is inherent in space and time.

Einstein didn't understand the physical nature of gravity. He only understood its effects as measured in terms of space-time curvature.

Einstein didn't understand the nature of time and space. He only understood that the laws of nature were conserved across whatever changes took place in an object's behavior in relation to measurements of time and space, and this dimension of the invariance of law in physical events was the central principle that tied special and general relativity together.

If Einstein didn't understand the nature of gravity, and if he didn't understand the nature of space and time, then Einstein certainly did not understand the nature of the dynamic – if any – among gravity, space and time. What Einstein did understand was the manner in which the curvature of the space-time metric could be used to describe the way in which the presence of gravity could affect the world line trajectories of objects in relation to that presence.

Several generations of scientists (along with the general public) have been led to believe that Einstein showed how gravity affects/warps time and space. He did nothing of the kind, but Einstein did advance the ability of physicists to more accurately describe, under a variety of conditions, the behavior of objects in gravitational fields, along with the properties of such fields.

Einstein theory of general relativity did not overthrow Newton's conception of space and time. Notwithstanding the foregoing statement, the opening sentence of this paragraph does not imply that Newton's understanding of space and time was correct or that Einstein's ideas concerning the malleability of space and time are incorrect ... rather, the initial claim of this paragraph indicates that Einstein did not demonstrate that gravity is capable of warping either space or time and, therefore, did not demonstrate that time and space were malleable in some fashion.

Einstein's theory of general relativity overthrew Newton's way of methodologically engaging gravitational behavior. The metric of description used in general relativity (namely, space-time), along with the central equation of general relativity (that showed how gravitational effects could be mathematically represented in terms of such a metric) were at the heart of Einstein's conceptual revolution ... the revolution was one of method not ontology.

A little over a century and a half after Newton published *Principia Mathematica* (which contained his theory of universal gravitation), the heuristic value of Newton's work was once again in evidence. In 1846, Urbain Le Verrier, a French mathematician who was interested in celestial mechanics, performed various Newtonian-based calculations in conjunction with some anomalies that were present in the orbit of Uranus, and he concluded that the planet's orbital oddities might be due to the presence of an unseen, cosmic neighbor.

Le Verrier didn't have to wait long to learn the fate of his conjecture. A few months after he made his claim, German astronomers located the unseen neighbor – now known as Neptune – moving along its orbit in accordance with Newtonian physics.

Over a decade later, Le Verrier discovered an anomaly in the orbital behavior of another planet. This time the object of interest was Mercury.

Mercury's perihelion (the point of closest approach to the Sun) differed from what Newtonian physics predicted it should be. The difference between reality and prediction was about a half a second, and, as a result, Mercury's precession (the movement of the planet's axis as it spin's about the Sun) was slower than expected.

Some individuals wondered if the orbital anomalies of Mercury were due, like those of Uranus, to some unseen object that was affecting the way in which the axis of Mercury moved as the planet orbited the Sun. This invisible possibility was given the name of Vulcan, and astronomers went in search of its existence ... to no avail.

The reason why astronomers searched in vain for Vulcan is because it did not exist. Instead, the cause of the anomalies in the precession of Mercury at perihelion was due to a calculating glitch of sorts.

Newton's theory of gravity couldn't produce the right answer. Einstein's theory of general relativity could.

The difference in accuracy between the two theories amounted to a mere millionth of a percent. This difference in accuracy between Einstein and Newton showed up other contexts as well.

For example, calculations have been made in relation to certain kinds of pulsars (believed to be the collapsed remains of very large stars) known as binary pulsars ... that is, pulsars that have a close relationship with one another. The perihelion precessions of such pulsars were reflected more accurately by Einstein's theory of general relativity than by Newton's universal theory of gravitation.

The anomaly in the precession of Mercury was said to be due to the way that the Sun warped space-time in the vicinity of Mercury. This curvature affected the world line of Mercury as it moved along its orbit.

Similarly, the precession anomalies associated with various binary pulsars is due to the way in which space-time allegedly is warped in the vicinity of those cosmic objects. Remember, however, that the warping of space-time in the case of binary pulsars (as is also the case in relation to Mercury) gives expression to the way in which the underlying metric for describing curvature by means of the field equations of general relativity reflects the behavioral properties of the gravitational phenomena that are taking place and, consequently, neither space nor time is necessarily being warped nor curved.

Another prediction of – and, therefore, test for – general relativity involves the way light passes through a gravitational field. According to Einstein, the presence of massive bodies – such as the Sun – should introduce a curve into the way light journeyed through the gravitational field generated by such bodies.

On May 29, 1919 the sun was scheduled to undergo an eclipse. At that time, the Sun's path along its orbit would carry it across some background stars that were known as the Hyades.

Since the eclipse would cover up the luminosity of the Sun that, normally, would block visibility of the background stars, observers would be able to see what happens, if anything, to the light from those objects. If the general theory of relativity were correct, then, the light that is traveling through the gravitational field of the Sun from the background stars should be deflected to some degree.

Arthur Eddington organized an experimental field trip in which several points of observation were established to view the eclipse against the background starlight. One point of observation was on an island (Principe) off the west coast of Africa, while the other post was set up in Brazil (Sobral).

There is some indication that Eddington might have cherry-picked the data generated by his experiment and framed his results in accordance with such selective use of data. In any event, Eddington claimed that his expedition had vindicated Einstein because the gravitational field of the Sun deflected the background starlight in just the manner that the theory of general relativity predicted.

Whatever truth might, or might not, be present in the foregoing allusion to the possible manipulation of data by Eddington, the issue is neither here nor there as far as the bending of light is concerned. This is because there have been many other studies verifying that light is, indeed, bent in the presence of a gravitational field that is sufficiently strong, and the phenomenon is referred to as "gravitational lensing".

Again, one must distinguish between the ontology of a gravitational field and the metric that is being used to model that field. The latter model is rooted in the metric of space-time, and such a metric can be used to mirror the behavior of phenomena in a gravitational field … space-time actually reflects the properties of a gravitational field rather than either space or time

The warping or curvature of space-time in the model of general relativity describes the behavioral deformations that occur in a gravitational field under the right kinds of conditions. For example, with respect to gravitational lensing, light can be observed to bend in the presence of a gravitational field, but that bending has to do with the way in which the path of light is deflected in a gravitational field – a way that can be modeled through the general theory of relativity – and such bending does not have to do with the warping of space and/or time.

Although Einstein's theory talks in terms of the warping of space-time, neither space nor time is actually warped as light moves through a gravitational field. It is the gravitational field, itself, that contains whatever deformations are reflected in the model provided through general relativity.

Gravitational fields undergo warping and curvature. The world line of light describes the way in which such curvature affects the geodesic trajectory of light, but none of this necessarily has anything to do with changes in the ontological character of either space or time.

Einstein's ideas were put to the test again in 1959 at Harvard. Robert Pound, a physicist, and Glen Rebka, Jr., a grad student, devised an experiment to test whether, or not light – as general relativity predicted – would display a shift in its frequency (toward the red end of the spectrum) in the presence of a gravitational field of suitable strength.

Pound and Rebka, Jr. used an elevator shaft in the Jefferson laboratory building to conduct their experiment. In the basement portion of the elevator shaft, they placed a radioactive isotope of iron that would radiate gamma rays toward, among other places, the roof of the building.

On the roof of the building was an instrument that could detect the presence of gamma rays. Although the distance from basement to roof was 73.8 feet, the two researchers observed a drop in the energy of the gamma rays as the radiation made its way from the bottom of the shaft to the top of that structure.

The drop in energy was due to the impact that the Earth's gravitational field had on those gamma rays as they sped toward the top of the building. Following the shortest route (geodesic) through the curvatures/warps present in the Earth's gravitational field, the gamma rays lose a few trillionths of a percent in relation to the energy with which they began their trip.

The effect of such a loss in energy stretches the wavelength of the gamma radiation. The phenomenon is known as "gravitational redshifting".

The effect is small. However, as Pound and Rebka, Jr. demonstrated, it exists.

The foregoing issue has some relevance to the discussion that took place in 'Chapter Two: The Meaning of Red'. In astronomy, shifts toward the red end of the spectrum are often interpreted to mean that light undergoing such a red shift must be receding from Earth, however, as Pound and Rebka, Jr. have shown, shifts toward the red end of the spectrum also give expression to gravitational redshifting.

Obviously, the effects of gravitational redshifting will have to be taken into consideration when trying to figure out what meaning should be given to wavelengths arriving from the cosmos that have been redshifted. Such considerations will involve an estimate of how many different kinds of gravitational fields light travels through on its way to Earth as far as the strength of such fields are concerned, and as far as how long light is under the influence of those fields is concerned.

If the wavelength of gamma radiation can reflect a lost in energy over just 73.8 feet due to the influence of the Earth's gravitational field, what happens to the wavelength of light that is subjected to much stronger gravitational fields for much longer distances (and, therefore, greater periods of time)? The phenomenon of gravitational redshifting, along with the previously discussed findings of Halton Arp (who claims that redshifting has something to do with the birth of new galaxies and not an expanding universe), introduces additional layers of complexity into the problem of trying to establish the meaning of red shifts that are observed in the wavelength of the light that is being received from different parts of the universe.

According to many physicists, just as the presence of gravity stretches the wavelength of, say, light, gravity also supposedly impacts time. For example, in 1976, NASA conducted 'Gravity Probe A' that involved comparing an atomic clock on Earth with one that was aboard a space vehicle some 6000 miles above Earth.

The atomic clock stationed above the Earth ran slightly faster than the atomic clock on Earth. The difference was attributed to the impact gravity has on time ... namely, that clocks run faster in weaker gravitational fields than they do in stronger gravitational fields.

The differences were small ... around 70 parts per million. However, Einstein's general theory of relativity was able to describe the situation very precisely.

In 2011, researchers at NIST (National Institute of Standards and Technology) demonstrated that even differences in elevation as small as one foot had an impact on the running of clocks. The more elevated clock (by a foot) operated at a slightly faster rate than did the clock that was one foot lower.

Just as Einstein was mistaken when he stated that time is what a clock measures in conjunction with special relativity, so too, one makes a mistake when one tries to argue in general relativity that differences in the rate at which clocks run in different kinds of gravitational fields has anything to do with the nature of time. Impacting the rate at which a clock runs is about the way such mechanisms interact with a gravitational field, and, therefore, that dynamic carries no implications concerning the ontology of time.

Not too long after Pound and Rebka, Jr., performed their 1959 gravitational redshifting experiment at Harvard, scientists began to think about another way to test the general theory of relativity. This challenge involved the idea that, according to Einstein, massive bodies supposedly pull space-time along with them as those bodies move along their orbits.

Thinking of a possible experiment is one thing. Having the technological and financial means to perform such an experiment is often quite another matter altogether.

Nearly half a century after conceiving of an experimental test for the general theory of relativity in conjunction with the prediction that the gravitational field of a body pulls or drags space-time along with that body as the latter moves and spins along its orbit, both the technological and financial wherewithal came into alignment to bring the original idea to fruition. More specifically, gyroscopes were placed on a satellite set to orbit Earth, and as a way of establishing a framework for comparing results relative to the gravitational field of the Earth a point of reference (a star) was selected for aligning both the gyroscope and the satellite.

Scientists began to search for any changes in the alignment of the gyroscopes. Such changes, if discovered, meant that the gyroscopes alignment was being altered due to the presence of the Earth's gravitational field ... in other words, the gravitational field of the spinning Earth was dragging the gyroscopes along as the Earth moved along its orbital path.

Results of the foregoing experiment were released in 2011. The experimental data indicated that the Earth's spin was dragging space-time along with it, and this was reflected in the changes in alignment in the gyroscopes placed on the satellite orbiting Earth ... changes that were fully in accord with the theory of general relativity.

The gyroscope experiment does not actually demonstrate that the gravitational filed associated with the Earth's spin drags space-time along with that movement. The foregoing experiment demonstrates that the gravitational field associated with the Earth's spin has ramifications for what happens to gyroscopes under such conditions.

The model to which general relativity gives expression involves a metric known as space-time that will display deformations or curvatures that are functionally related to the description of the world line or trajectory of gyroscopes whose measurements are embedded in a metric whose behavioral properties are descriptively shaped by the Earth as the latter spins and moves along in its own gravitational field. Data points are collected that describe how the alignments of the gyroscopes have changed in relation to the movement of the Earth.

The Earth is not dragging space and time as the planet spins and moves along its orbital path. However, changes in space-time configurations -- which constitutes a representational model of gravitational phenomena -- indicate that the metric of space-time has been deformed (i.e., dragged) within the framework of the general relativity model, and this deformation is a descriptive reflection of the way the gravitational field of the Earth behaviorally interacts with the gyroscopes (and vice versa) aboard the satellite.

The spinning, orbiting properties of the Earth drag along portions of the surrounding gravitational field, and the foregoing dynamic is captured during the process of plotting various world lines within the space-time metric. Neither space, nor time, nor space and time are getting dragged along.

The general theory of relativity was tested in, yet, another way. The theory predicted that when dense, massive objects rotate about one another, they should create waves in the surrounding gravitational field.

The waves were interpreted as being the result of the way massive, fast moving bodies pinch space-time, creating ripples in that fabric. However, one could just as easily interpret the foregoing ripples or gravitational waves as being a manifestation of the gravitational field itself rather than being ripples in space and time per se.

The size of such gravitational waves is quite small. Nonetheless, over time, those waves would drain energy from the system and, in the process, cause the respective orbitals to deteriorate and induce the two bodies to spiral into one another.

In 1974, Russell Hulse and his thesis advisor, Joseph Taylor, Jr., discovered a binary pulsar. Its official name reverberated liltingly off the tongue as: PSR B1913 +16, but, despite such tonal attractions, the cosmic system is sometimes referred to as the Hulse-Taylor binary pulsar.

The binary system consisted of a pulsar and a black, or visually unseen, companion object (which is believed to be an ultradense neutron star). Emanating from the system are pulses of very regular, stable radio waves that scientists consider to be linked to the rotation of the neutron star that constitutes the pulsar portion of the binary system.

Five years later, the researchers detected the presence of small acceleration effects in the orbital behavior of the foregoing pulsar. This finding was considered to be an indication that gravitational waves were being generated by the system.

The Hulse-Taylor binary pair has been studied for more than three decades, and measurements indicate that the two members of that system are, indeed, spiraling in toward one another. The rate of that spiraling process conforms to the predictions of the general theory of relativity and, therefore, constitutes a confirmation of Einstein's ideas.

Inferring the presence of gravitational waves is one thing. Detecting their presence is quite another matter, and indeed, Einstein believed that the size of gravitational waves is so small that he doubted whether waves that size could ever be detected.

Despite Einstein's misgivings, some detection projects are already up and running. For example, consider the half-billion dollar LIGO project.

LIGO stands for: Laser Interferometry Gravitational Wave Observatory. Laser interferometry refers to the process of sending laser beams down two, perpendicular arms and, then, measuring any differences in length between the two.

If gravitational waves are large enough, they will affect LIGO. Such waves supposedly will alter the relative length of the arms as the waves ripple through the instrument system.

LIGO is an Earth-bound project. Two of its three components reside in Washington State, while the other instrument package is based in Louisiana.

All three of the components of LIGO are precision instruments. They can detect movements that are no bigger than the diameter of an atom.

There is a spaced-based version of LIGO, named LISA (Laser Interferometer Space Antenna) that has been on the drawing boards for years at the NASA Goddard Space Flight Center and the European Space Agency. If it becomes operational, it will consist of three satellites or spacecrafts that form an equilateral triangle whose sides are 3.1 million miles long, and those distances will be monitored continuously in an attempt to detect the presence of any gravitational wave that ripples through any portion of the triangle and, in the process, affects the lengths of the sides of the space-based equilateral triangle.

Another approach to the detection of gravitational waves involves something called atom interferometry. Like LISA, this possibility would also employ three satellites or spacecrafts, but instead of measuring changes in laser signals, atom interferometry measures changes between atom clouds located just outside of a spacecraft rather than between distant spacecrafts.

Atom interferometry has one advantage over laser interferometry. The former process tends to be more precise than the latter method.

Irrespective of the issue of precision, currently, both LISA and the atom interferometry possibilities are so much pie-in-the-sky. At the present time, each of those approaches lacks the necessary funding to move forward.

Using gravitational waves to probe the cosmos has a potential that is not enjoyed by the use of electromagnetiism. More specifically, if the Big Bang theory of cosmic origins is correct, the manner in which the first 380,000 years of the universe unfolded led to the trapping of photons in a light-absorbing particle/radiation soup and, thereby, formed a curtain that prevents anyone from penetrating that curtain (through photon-based modes of detection) and looking (again, in terms of photon-based processes of seeing) at what happened on the other side (temporally speaking) of the drawn curtain.

However, no such cosmic curtain exists in relation to gravity, and, consequently, one might be able to use the effects of gravity (such as in the form of gravitational waves) to develop a sense of what might have been transpiring in the very early universe (just a billionth of a billionth of a billionth of a billionth of a second after the Big Bang). In addition, gravitational effects – such as waves -- might permit scientists to probe what transpired at various points before the Cosmic Microwave Background Radiation became visible (some 380,000 years after the Big Bang) as electrons decoupled from the particle/radiation soup and began to form atoms as they hooked up with the positive ions that were present in that soup.

One process through which gravitational waves might have been generated in the early universe is if substantial amounts of matter were moved about very quickly. Moreover, one way to bring about such movement might have been through a phase transition of some kind.

For example, proponents of the Big Bang believe plasmas made of quarks and gluons existed very shortly (extremely shortly) after the initial explosive event. As the temperature of the early universe cooled slightly, there was a change in phase state at some point, and when this occurred, the quark-gluon plasma was replaced by protons and neutrons.

According to some astrophysicists (for example, David Spergel) bubbles arise in first-order phase transitions and begin to collide with one another in a violent fashion until the old phase is completely replaced – more or less – by the new phase state. Such scientists believe that the collisions that take place during this sort of phase transition would have generated gravitational waves (some physicists have argued that the turning on the Higgs field in the early universe also might have constituted the sort of phase transition that could have led to, or been involved with, the generation of gravitational waves).

Another possibility that might have led to the creation of gravitational waves in the early universe involves quantum fluctuations. Such fluctuations – so the theory goes – would have caused some regions of space-time to expand more than other regions, and in the process would have generated ripples that are referred to as "stochastic gravitational waves".

Detecting such waves – if they exist -- is a very delicate process. One would have to have instruments – and modes of analysis -- capable of differentiating between stochastic gravitational waves and other possible sources of gravitational waves [e.g., those that are generated through the aforementioned phase transitions or caused by Hulse-Taylor-like binary systems or which might arise in conjunction with the collision of galaxies or that might be associated with supernovae or arise in conjunction with the activity of black holes (if they exist)].

In March of 2014, an announcement was made (in relation to the BICEP2 program) claiming that the detection of primordial gravitational waves – along the lines indicated during the last three paragraphs – had been achieved. The group of scientists making the announcement had been working with a microwave telescope that was located at the South Pole.

The method of detection involved in the alleged foregoing discovery revolved about measurements of polarization in the Cosmic Microwave Background Radiation. Theoretically, gravitational waves would show up by underwriting amplitudes of greater oscillation in the CMB along one line of measurement compared with what has happening along some direction that was perpendicular to the first line of measurement.

Unfortunately, there can be more than one source for whatever polarization that might be detected in conjunction with Cosmic Microwave Background Radiation. The polarized dust that is present in the galaxy is one source of such contamination.

There is a potential way out of the foregoing problem because there appears to be a difference between the modality of polarization that is associated with gravitational waves and what is found in relation to other possible sources of polarization. This potential way of resolving things revolves around whether, or not, there is a certain kind of twisting pattern in the polarization ... gravitational waves purportedly possess such a pattern and are referred to as B modes, but the aforementioned twisting pattern is absent in other forms of polarization that are referred to as E modes.

Gravitational wave researchers also make reference to a ratio when discussing their results. More specifically, the ratio – denoted by the letter 'r' – relates the presence of a possible gravitational wave signal to the magnitude of temperature fluctuation signals known to be present in Cosmic Microwave Background Radiation through previous measurements.

According to the European Space Agency (and based on the measurements carried out by its Planck satellite) the value of 'r' can range from: Zero, indicating that no gravitational waves appear to be present, to: An upper parameter of 0.13. Anything in between these two boundary markers suggests the presence of gravitational waves.

The value of 'r' reported by the researchers at the South Pole in March of 2014 was 0.2. This was well above the boundary value for 'r' that had been cited by the European Space Agency.

As always occurs when someone makes an announcement concerning a discovery – or, at least, this should occur – various scientists and researchers have gone back through the original data and reanalyzed it by taking into consideration different possibilities, new research, and alternative methods that might cast a different, critical light on the initial discovery. The results of that critical analysis indicates that all, or a considerable, portion of the foregoing 'r' value found by the BICEP2 researchers might be attributable to the polarizing potential of galactic dust.

For instance, since the March/2014 BICEP2 announcement, the Planck satellite has generated date indicating there might be much more dust in the Milky Way than was assumed by the members of BICEP2. In general terms, polarized dust from our own galaxy is very difficult to eliminate as a possible source of contamination when calculating values of 'r'.

Thus, the jury is still out as to whether, or not, gravitational waves actually have been detected. Someone will have to confirm the foregoing results in a way that strongly indicates that the 'r' value is due to the presence of gravitational waves rather than due to other possible sources of polarization.

However, even if BICEP2 did successfully detect the presence of gravitational waves, and even if some other research team is able to independently confirm the BICEP2 findings, can one necessarily conclude that a transition in phase state or some sort of quantum fluctuation that occurred in the early universe generated the gravitational waves being detected.

Just as polarization can have a variety of sources, so too, gravitational waves can be generated in a variety of ways ... some of which are known and some of which might not, yet, be known. For example, could activity involving Dark Matter -- in relation to either the early universe or at some other point in cosmic time -- be responsible for the generation of detectable gravitational waves or could the collision of black holes give rise to detectable gravitational waves? Or, maybe, the pinching effects associated with intense and powerful electromagnetic activity in the plasmas that occupy large portions of the universe might also be able to set gravitational waves in motion.

How would one differentiate between the gravitational waves generated by, say, a phase transition or quantum fluctuations in the early universe and gravitational waves that were generated through some of the other possibilities suggested in the last paragraph? This question becomes especially acute when one reflects on the following possibility: If the Big Bang did not occur, then, one is going to have to re-conceptualize the significance of whatever gravitational waves might be detected, or, alternatively, even if the Big Bang did occur, it is possible that something like Dark Matter – if it exists – could be responsible for the generation of gravitational waves that, currently, might be interpreted as being due to some other cause ... such as phase transitions of one kind or another or such as quantum fluctuations ... fluctuations that -- as has been pointed earlier in this volume as well as in *Volume II* of the *Final Jeopardy* series -- might not be as ubiquitous as many quantum theorists assume.

Further tests of general relativity are being devised. For example, one project – known as STEP (Satellite Test of the Equivalence Principle) – involves (as the parenthetical expression indicates) probing one of the central tenets of general relativity – namely, the equivalence principle.

Among other things, the equivalence principle means that different bodies, quite independently of their composition and mass, will behave the same way when subjected to one and the same gravitational field and all other possible influences (e.g., air resistance) are eliminated. Numerous experiments have been performed on Earth confirming the equivalence principle with a precision down to one part in ten trillion.

However, there are many possible sources of Earthly contamination (e.g., vibrational noise) that can affect the results of experiments involving the equivalence principle. By shifting the laboratory to space, one can eliminate such sources of contamination and increase the accuracy of measurements by as much as a factor of $10^5$.

All theoretical attempts to unify gravity with the other three fundamental forces entail violations of the equivalence principle. In view of the precision with which the equivalence has been measured already, if violations of the equivalence principle are possible, then, one likely will need a measurement process that is extremely sensitive to the presence of clues that might be very subtle and difficult to detect.

The aforementioned STEP project plans to use four pairs of experimental masses that are composed of very different materials. The materials will be cooled to just a few degrees above absolute zero in order to prevent even slight fluctuations in temperature from contaminating the measurement process.

Next, the test masses will be subjected to some version of free fall. STEP will be looking – in very precise ways -- for indications that materials of different composition fall at different rates and, thereby, give expression to a violation of the equivalence principle.

Whether, or not, STEP will ever become realized is, so to speak, up in the air. As is the case with many scientific projects, there are only so many research dollars to go around, and, consequently, not every project will be funded or, if funded, such projects will not necessarily receive adequate funding.

Conceivably, even if projects like STEP get off the ground and are permitted to put their experimental design to the test, there is no guarantee that subtle violations of the equivalence principle will be discovered. As noted previously, the currently available theoretical models that are attempting to unify gravity with the other three forces all indicate that, at some point, the equivalence principle will be violated, but such models might not be correct, and, if this is the case, then how does one interpret experimental outcomes that fail to detect any violation of the equivalence principle?

Perhaps, null results might mean that one needs to select materials of different composition to be tested. Or, possibly, a failure to detect violations of the equivalence principle just means one needs to improve the accuracy of the measurement process.

On the other hand, perhaps such results indicate that one needs a new model of unification ... that our current models are not capable of correctly capturing the conditions under which the equivalence principle will be violated (if it can be). Or, maybe null experimental outcomes allude to the possibility that the equivalence principle might prove to be inviolate no matter how precise and subtle one's experimental wherewithal might be.

The idea of unification is a theoretical ambition. If the fundamental forces resist all of our tests to prove some theory of unification concerning the fundamental forces of nature, this would not be the first time that theoretical ambitions (irrespective of how beautiful and alluring they might be) will have flown into the jagged edges of inconvenient facts.

At the very least, Einstein devised a model that, among other things, permits one to calculate more accurate answers to various problems in physics than can be accomplished through Newton's universal theory of gravitation. Nonetheless, in doing so, Einstein did not necessarily bring the world of science any closer to understanding what gravity is or what makes it possible.

General relativity is about the behavior of gravitational fields. It says nothing about how gravity makes such fields possible.

The tensors present in the field equations of general relativity (tensors involve a method of geometric description that allows one to analyze the linear relationships among various vector, scalar and tensor forces) might permit one to better explore the dimensions of curvature that are present in gravitational fields being represented through a space-time metric than can be accomplished through the methods of calculus that are used to analyze the three-dimensional gravitational fields that are engaged through Newtonian theory. Nonetheless, neither of the foregoing two theories of gravity is capable of penetrating to the actual nature of the source of gravity.

Einstein's general theory of relativity is a background independent model. By distancing himself from the absolute conceptions of space and time that formed the backdrop against which Newtonian physics played out, Einstein made 'space' and 'time' functions of his field equations. This is why, in general relativity, space and time are both considered to be malleable in accordance with the requirements of the theory's equations involving tensor curvature.

By making general relativity background independent with respect to space and time, Einstein established a geometric framework (the tensor curvature of space-time) by means of which one could make calculations concerning the degree to which the force of gravity is present at any given point in such fields under various conditions of interaction. In doing so, Einstein laid the seeds of confusion concerning the ontological significance of events in space-time.

From the perspective of general relativity, changes in the geometric configuration of gravitational fields being described through the tensor curvature manifested in the metric of space-time were construed as having something to do with the ontology of space and time. And, yet, if general relativity is truly background independent, then, changes taking place within the space-time metric really don't necessarily carry any implications for either space or time.

Quite soon after the general theory of relativity was released, Einstein realized that one of the implications of his theory was the existence of gravity waves. He also understood that those waves carried energy and that if general relativity were going to be reconciled with quantum mechanics, then a way would have to be found in which quantum mechanics would be able to successfully describe how energy was manifested in, and distributed through, gravitational waves.

In 1929, Wolfgang Pauli and Werner Heisenberg were already busy reflecting on various possibilities concerning how to quantize fields of gravity. They felt that the issue might be dealt with in a way that was similar to the manner in which they had quantized electromagnetic fields in early versions of what came to be known as QED models (quantum electrodynamics).

However, one of the stumbling blocks with proceeding in the foregoing manner is the following: The medium for exchanging force in electromagnetism is the photon, and this particle does not does not interact with itself, but this principle does not hold true when it comes to gravity.

More specifically, the proposed medium of exchange for gravitational force – the graviton -- is inclined to interact with any source of energy, including itself, since all energy, due to its property of mass equivalence, possesses gravitational potential, and this dimension of self-interaction has proved very difficult to capture in a way that leads to sensible and relatively anomaly free solutions (e.g., this sort of self-interaction is replete with ghostly infinities that resist exorcism and, therefore, to date, has not been renormalized).

Over time, two different approaches emerged with respect to resolving the problems that surrounded the issue of how to handle or describe the energy of gravitational waves. One strategy sought to develop models that, like Einstein's theory of general relativity, strived to be background independent (and individuals such as Julian Barbour have gone even further than Einstein did in this respect).

The other strategy for resolving the problems associated with gravitational waves involved developing models that were background dependent. This led to various theories of quantum gravity, but due to, among other things, the infinities issue, all of those models floundered to varying degrees.

String theory arose, in part, as an attempt to turn quantum theory -- and, therefore, quantum gravity -- into a background independent framework. One of the many attractive features of such a background independent model is its capacity to resolve the infinity problem since one-dimensional strings were capable of avoiding the anomalies that arose when one treated particles as dimensionless points.

Whatever the promise of strings might have been, that promise, to date, has not been realized. Despite all of the mathematical bells and whistles that have become associated with string theory (and such issues have been explored somewhat in Chapter 5 of *Final Jeopardy: Physics and The Reality Problem Volume II*), no one knows how to identify the version of string theory that applies to the observable universe, and, in addition, no one knows how to test such a theory even if such a theory were identifiable because the scale on which string theory operates is far beyond our capacity to experimentally probe ... in short, string theory entails a wealth of mathematical possibilities but offers very little in the way of tangible science.

Quantum gravity deals with events on the level of the Planck length that carries a value of $1.616199 \times 10^{-35}$ meters. The foregoing figure is surprisingly precise for something that might not have any actual physical significance.

Physicists speculate about what might take place on the level of the Planck length. They have to speculate because the Planck length is many, many, many orders of magnitude beyond what any mode of current measurement is capable of exploring or testing.

Some physicists believe that the Planck length constitutes the shortest possible measureable "space". Irrespective of whether, or not, they are correct in that assessment, the nature of ontology is not a function of whatever limits might exist in conjunction with the process of measurement, and, therefore, there is nothing in principle (at least that we know of) that prevents ontological size from being able to extend down to the level of, say, $10^{-100}$ (a googol) or to even more unimaginable realms of smallness.

Ontology is what it is. Our capacity to measure the nature of that ontology is an entirely separate matter.

Some physicists have speculated about what "space" might look like at the Planck length. Indeed, various individuals believe that space itself could be quantized at the level of the Planck length ... although such speculations appear to be somewhat arbitrary since they tend to be based on little more than the needs of the theories that give expression to such speculations.

Earlier in the present book, a question was raised about whether, or not, space is 'something' that is capable of being inflated. Similarly, one can now ask whether, or not, space is something that is capable of being quantized?

We don't know whether, or not, space can be inflated. Furthermore, we don't know whether, or not, space can be quantized.

Does space interact with the objects it contains? We don't know, and even if space did interact with such objects, how would we go about detecting the presence of that dynamic since we don't know what space is, and, therefore, we don't know what the character of the dynamic is to which space gives expression and for which we should be looking for signs involving the presence of that kind of activity.

The notion of the "Planck Scale" is another way of alluding to what might be transpiring within the realm of quantum gravity. This is an energy scale whose value is: $1.22 \times 10^{19}$ GeV (billion volts) … an energy scale that is many orders of magnitude beyond our current capabilities to generate [the revamped LHC at CERN produces energies around 6-7 TeV (6000-7000 GeV or $6\text{-}7 \times 10^3$ GeV )], and, consequently, something that is not likely to be experimentally probed by humans any time in the near future.

The Planck scale of energy is considered to be important because such energies constitute the level at which the quantum effects of gravity supposedly begin to make their presence known. However, this is also the level at which quantum field theories break down because of, among other things, the monstrous, non-renormalized infinities arising from the self-interaction of the gravitons that reside on that scale.

There is at least one person who has not been deterred by the difficulties surrounding peeking into the realm of the very small on the level of either the Planck length or the Planck scale of energy. His name is Craig Hogan, and he has the credentials and experience of a top-notch physicist.

Hogan does not believe space conforms to the traditional view of a smooth, continuous backdrop against which quantum events play out their dynamics. He believes that space is digital.

According to Hogan, the most fundamental stuff of the universe might not be matter or energy but bits of information. He believes the universe might arise from information.

Hogan, along with experimental physicist Aaron Chou, is building an instrument known as a Holometer. The instrument is intended to detect the noise or jitters of digital space.

The Holometer is an interferometer. It measures differences or changes (if present) between events transpiring along two perpendicular arms.

Laser beams will be sent down two sets of 40-foot long beam tubes that are under vacuum. Precision instruments allow the foregoing beams to be aligned, focused and analyzed for subtle changes in the length of the beams running down the different sets of beam tubes

Both quantum mechanics and general relativity tend to break down on the level in which Hogan and Chou are interested – that is, the level of the Planck length or Planck scale. This is a breakdown of description and methodology and not necessarily ontology.

Hogan maintains that the realm of the Planck length and Planck scale consists of information. Such information is stored on two-dimensional objects known as "light sheets", and that stored information is what becomes projected as the three-dimensional universe that we observe.

However, if Hogan is correct, there are three features of his model that are a little vague. (1) He doesn't account for how the light sheet initially came into existence or how information came to be coded on it. (2) He doesn't explain how the information on the light sheet came to be organized so that they gave expression to the physical laws and structures that make up the universe. (3) He is not able to explain how such information gets translated into becoming concrete structures that give expression to the dynamics of three-dimensional forms of physical energy, forces, and particles ... that is, the details of the projection process are a little sketchy.

Notwithstanding the foregoing considerations, Hogan believes that the first step to take toward establishing the possibility of a holographic universe is to determine whether, or not, space is digital in character. In order for Hogan's version of the holographic principle to be possible, space has to be shown to be bit-like in nature.

The Holometer experimental apparatus is intended to test the foregoing idea. More specifically, by detecting changes (if they occur) in the length of the signals that have been sent down different arms of his two-pronged Holometer/interferometer, he believes he will generate data that confirms the digital character of space since, according to Hogan, those changes will be the result of the way that the digital jitters of the bits that make up space have jostled the interferometer.

The sort of jostling that Hogan and Chou are searching for has a specific frequency. A million times a second is the rate at which the to and fro of spatial jitters supposedly takes place.

What would cause space to jitter at such a frequency is unknown. As a result, the choice of frequency seems rather arbitrary in character, and if space is digital, what is the "medium" in which space jitters take place?

Even if the two aforementioned researchers were to detect the presence of such a frequency, that sort of experimental determination does not necessarily confirm their underlying idea about the digital nature of space. For example, in light of the fact there are many aspects of the universe that fall beyond the horizons of our understanding, why should one automatically assume that the discovery of a frequency that occurs at a rate of a million times per second constitutes evidence for the presence of the jitters that supposedly are inherent in digital space?

Why assume that space, even if digital in character, jitters at a frequency of a million times a second? More importantly, why suppose that space, even if digital in character, is subject to the jitters?

Finally, even if digital space does have jitters that exhibit a frequency of a million times a second, none of this resolves any of the issues noted earlier. That is, even if one were to demonstrate that space is digital in nature, this does account for how light sheets form and, then, become encoded with the laws of physics that, in turn, are holographically projected onto existence as a three-dimensional universe.

Chapter 6: Mysterious Holes

Karl Schwarzchild caught a glimpse of them within the field equations of the theory of general relativity soon after the latter was made public. Notwithstanding the apparent implications of his own theory, Einstein didn't believe they would ever become a reality. Today, many astrophysicists contend that – despite never having seen them – there are supermassive versions of them buried within the heart of many galaxies (at least one in ten).

What is the mysterious entity to which Schwarzchild, Einstein, and modern astrophysicists are alluding? They are referring to what eventually came to be known as "black holes" … a term coined by the physicist, John Wheeler.

While the standard theory of the universe holds that stars share many features in common, various kinds of stars follow life cycles that differ in certain ways. For example, according to the standard model of star formation, all stars begin in clouds of gas and dust, but over time, such materials become condensed through the influence of gravitational attraction.

When such proto-stars have become sufficiently massive, they ignite, starting a process of nucleosynthesis. That is, through a process of fusion, more and more complex atoms form … a process that starts with hydrogen and ends, millions of years later, with iron.

According to mainstream astrophysicists, stars avoid gravitational collapse through the outward pressure that is generated by the fusion process. Such pressure serves as a countervailing force to gravitational attraction.

When all of a star's nuclear fuel has been converted to iron, the process of fusion stops because iron in incapable of continuing the process. When fusion stops, there is no longer any pressure generated by that process to counteract the force of gravity, and, as a result, the core of the star begins to collapse.

As the size of the compacting core reaches approximately the size of Earth, the phenomenon of electron degeneracy assumes importance. This phenomenon involves the creation of countervailing pressure by fast moving electrons and, for a time, this prevents further gravitational collapse from occurring.

Electron degeneracy is said to be an application of the Pauli exclusion principle in which no two electrons can occupy the same state at the same time – even under the crushing pressure of gravitational collapse. However, what actually makes such a dynamics operative is not fully understood.

At this point, stars that are less massive than 1.39 times the size of our Sun become white dwarfs. This is known as the Chandrasekhar limit.

Those stars, on the other hand, that are, somewhat, more massive than the Chandrasekhar limit continue to collapse further because whatever amount of electron degeneracy is present is insufficient to prevent that collapse. One wonders what happens to the Pauli exclusion principle under such circumstances.

Gravitational collapse continues until it is resisted by a pressure that is created through neutron degeneracy. Although neutrons carry no net electrical charge, they tend to repel one another under certain circumstances, and this seems to give expression to a variation on the Pauli exclusion principle.

Neutron stars are formed in the process. By this time, the star has collapsed to a size that is about 6.2 miles in diameter (one wonders what happened to the tendency of neutrons to repel one another, and this issue has led some scientists to wonder if neutron stars are even possible.).

Stars more massive than the ones that supposedly end up as neutron stars have another fate waiting in store for them. In such cases, continuing collapse cannot be prevented by either electron or neutron degeneracy.

According to the general theory of relativity – at least as interpreted by individuals such as Karl Schwarzschild – the next way station for such a star is when a singularity is established at the star's core and a black hole arises. At this point, physics breaks down and is unable to describe what is taking place within a singularity.

According to Einstein's general theory of relativity (at least as understood by Karl Schwarzschild), if a given region of space-time became sufficiently deformed by the presence of a compact, massive source of gravity, then, that region would be characterized by a number of properties. Included among those properties is the following feature: When a given region of the cosmos acquires a gravitational attraction that is sufficiently powerful, then, neither particles nor various forms of electromagnetic radiation would be able to escape from such a region, and, therefore, that portion of space would not give off light.

Schwarzchild was not the first individual who envisioned such a possibility. John Michell, an 18th century clergyman and natural philosopher -- who had devised the instrument used by Cavendish to experimentally determine the mass of Earth (and, indirectly, the value of the gravitational constant) -- conceived of such a possibility around 1783 and referred to them as "dark stars". Dark stars are stellar objects whose gravitational attraction is so great that not even light could escape from their grasp.

Schwarzchild wrote his groundbreaking paper on general relativity while serving on the Russian Front during World War I with the German military. Shortly, thereafter, he died ... apparently from an autoimmune disease rather than from some instrument of war.

For more than four decades, Schwarzchild's ideas were, in many respects, as deeply buried as his body was. However, beginning with the work of David Finkelstein in 1958, interest began to pick up in conjunction with Schwarzchild's solution to the field equations of general relativity and such interest fertilized the growth of a very active area of theoretical research that has captured the imagination of even the general public through the work of, among others, Stephen Hawking.

The theory of general relativity is expressed in terms of differential equations that analyze the rate at which different parameters of the theory change under various conditions. Schwarzchild's mathematical work in relation to general relativity provided the first solutions for such equations.

Imagine there is a non-rotating, concentrated, spherical mass, M, with radius R. In order for an object of mass, m, to be able to escape the gravitational attraction of the spherical mass, M, the mass, m, will have to have a certain kinetic energy of motion – namely, $\frac{1}{2}mv^2$ (where 'v' is the velocity of m) -- that gives expression to an escape velocity greater than the gravitational attraction that is exerted by the spherical mass, M on the small mass, m -- namely, GMm/R.

If GMm/R is greater than ½mv², then, m will not be able to achieve escape velocity. This will be true even if the velocity of v turns out to be the speed of light, c.

If one solves for R (and considers 'v' to be 'c', the speed of light) one arrives at: $2GM/c^2$. Thus, if the radius, R, of the spherical mass, M, is $2GM/c^2$, then, not even light can escape from such a concentrated mass.

$2GM/c^2$ is referred to as the "Schwarzchild radius". The outer surface of a spherical mass with such a radius is called the "event horizon".

The concentrated, spherical masses described through Schwarzchild's mathematical engagement of general relativity are non-rotating. These are considered to be the simplest versions of black holes, consisting of just a non-rotating core, or singularity, and an event horizon. This is referred to as the Schwarzchild type of black hole.

Black holes that have a rotating core are known as Kerr type black holes (based on the work of Roy Kerr in 1963 and 1965). The core of such black holes conserves the angular momentum of the rotating star from which such black holes arose.

Kerr black holes are further subdivided. There are charged (Kerr-Newman type) and non-charged (Kerr type) editions of rotating black holes.

Unlike black holes of the Schwarzchild type, Kerr black holes, have a region that is known as the ergosphere. This region is considered to be egg-shaped and is formed through the way in which the rotation of the black hole is believed to gravitationally drag the space that borders on the black hole's event horizon.

Theoretically, an object could enter into the region of a Kerr type black hole's ergosphere and manage to not only avoid being drawn past the event horizon, but, as well, to escape the seemingly inexorable gravitational attraction of the black hole altogether. This is possible because of energy that such a falling object might draw from the energy of rotation that is present in the ergosphere.

The term "singularity" is used to designate the heart of a black hole. Mathematically, the singularity consists of a point that is infinitely massive – at least, this is the case according to the mathematical calculations that are used to describe such an entity.

The singularity is that place or point where all mass in a black hole comes to its crushing conclusion. While the mathematics of black holes describe this core region in terms of being infinitesimally small, as well as being infinitely massive, no one really knows what is actually transpiring in such singularities.

Among other things, general relativity is not capable of dealing with the quantum effects that exist on the very small scale through which gravity supposedly imparts its force. Conceivably, but in some unknown fashion, such quantum effects might be responsible for preventing the force of gravity from becoming infinite in nature.

The presence of infinities is an indication that the underlying model contains anomalies of one kind or another. As indicated previously, known physics breaks down under the conditions of a singularity and, therefore, cannot reliably describe the nature or dynamics (if any) of that phenomenon.

Do quarks and leptons lose their identity "within" a singularity? Is there any "within" in a singularity?

Do forces – with, perhaps, the exception of gravitation -- disappear in relation to such a region? Is it possible that even the force of gravity is affected in some way beyond a certain point of collapse?

Does the Pauli exclusion principle cease to operate at the level of a singularity and, as a result, maybe leptons become boson-like in the former particles' willingness to associate with one another? Do emergent properties arise in singularities, and, in the process, generate strange phenomena that are comparable to (but different from), say, what goes on with superconductors or elements that exhibit the property of superfluidity.

For instance, is it possible that under extreme conditions of gravitational attraction something happens to the force of gravity and, in some way, attenuates or alters that force? Perhaps singularities give expression to some sort of super-plasma that, to some extent, counteracts the presence of gravity.

Whatever is, or is not, transpiring within a singularity, we are unable to extract any data from that entity. Like Las Vegas, what happens in a singularity, stays in the singularity ... unless ... .

There are a set of circumstances – at least theoretical ones – in which whatever is transpiring within a singularity might have ramifications for the neighborhood that is external to that singularity. Indeed, some theorists believe that powerful forces might be set in motion under the right kind of circumstances.

The term "naked singularity" is the phrase that is used to refer to the foregoing set of conditions. Naked singularities lack an event horizon.

Thus, there is no radius extending out from the core of such black holes to establish an outer perimeter that marks the point of no return. As a result, there might be nothing to prevent an object from approaching quite closely to such naked singularities and, yet, still be able to pull away.

Prior to mathematically working out the possibility for naked singularities in 1973, one influential perspective concerning event horizons was advanced by Roger Penrose in 1969. He maintained that every star that is sufficiently massive to collapse into a black hole necessarily generates an event horizon during the process of collapse.

Penrose's perspective was speculative in nature. According to his conjecture, the event horizon that formed during the collapse of a suitably massive star prevented the core singularity from directly communicating with the surrounding universe, and, therefore, his idea became known as the "cosmic censorship hypothesis".

In 1973, however, Hans Jürgen Seifert led a research group investigating the impact that the property of inhomogeneity (the different kinds of matter making up a star) might have on the process of black hole formation. Seifert and his research associates discovered solutions in which gravity did not become infinite and, therefore, the core of the black hole was not reduced to an infinitely strong gravitational singularity, and, as a result, the theory of general relativity did not break down.

More specifically, the numerical simulations run by Seifert and his colleagues indicated that the property of inhomogeneity seemed to lead to a layering of the matter that was being drawn toward the core of the collapsing star. This caused fleeting conditions of naked singularity.

Whether the foregoing conditions constitute a true naked singularity or merely gave expression to a transitional state in which the layers of inhomogeneous materials served as unstable, pseudo-like event horizons is not clear. In any event, various other researchers attempted to mathematically demonstrate that all naked singularities would be – relatively speaking – gravitationally weak (that is, not infinite in character), but they were unable to succeed in their attempts.

Subsequently, various researchers discovered the existence of solutions in which singularities were not gravitationally weak. These accounts involved the collapse of sufficiently massive, inhomogeneous, stellar objects that resulted in gravitationally strong singularities that were, nonetheless, naked.

In other words, the black holes that formed during the collapse of those stars lacked an event horizon. Nevertheless, according to relevant calculations such entities still would be visible to external observers.

Further numerical simulations gave expression to models that included parameters involving realistic values for gas pressure and density in inhomogeneous, sufficiently massive collapsing stars. These conditions also led to outcomes that involved naked singularities.

In addition to properties such as gas pressure and density, other parameters were introduced into the numerical simulations being explored. One of these parameters concerned the impact that the rotation of particles in a collapsing star might have on the nature of the resulting black hole, while another parameter involved the shape of the collapsing star ... most numerical simulations involved spherical stars, and, so, some individuals began to consider non-spherical possibilities.

In each of the foregoing cases – that is, rotating particles and non-spherical stars, solutions were found that gave expression to naked singularities. On the other hand, other parameters – such as gas pressure – were not simultaneously considered, and, therefore, whether numerical simulations that included parameters such as gas pressure, as well as the inclusion of non-spherical and rotational considerations, would have also led to naked singularities requires further examination.

All of the foregoing scenarios run contrary to Penrose's cosmic censorship hypothesis. That is, the foregoing numerical simulations indicate that event horizons do not always form when a suitably massive stellar object collapses.

The naked singularities that arise through numerical simulations of the foregoing kind cover an array of possibilities. In some situations, the condition of naked singularity is quite fleeting and, subsequently, becomes cloaked by an event horizon, whereas in other simulations, nakedness is a stable property of the collapsing stars being modeled.

The degree of inhomogeneity that is present in a suitably massive collapsing star can determine whether, or not, that collapsing star will become a traditional black hole (with an event horizon) or a naked singularity. If the amount of inhomogeneity is too little, then, a black hole will form, but above a critical threshold, inhomogeneity will lead to the formation of a naked singularity.

Similarly, the speed with which collapse takes place can affect what happens to a star that is sufficiently massive. If the speed of collapse falls below a certain threshold, then black holes are likely to form, whereas if the speed of collapse is above that threshold, a naked singularity is likely to form.

So far, naked singularities are nothing more than a function of numerical simulations involving a set of parameters involving such values as: gas pressure, density, degree of inhomogeneity, particle rotation, whether a star is spherical or non-spherical, the speed of collapse, and so on. Consequently, naked singularities might, or might not, exist.

To date, black holes also are hypothetical in nature. However, the degree to which black holes are hypothetical might be less speculative than the existence of naked singularities since there seems to be a certain amount of observation (more on this shortly) pointing toward the occurrence of phenomena that might begin to make sense if one were to consider black holes as real, rather than, as merely hypothetical entities.

Some astrophysicists believe that if naked singularities actually exist, they might serve as a laboratory through which to test various ideas in string theory and loop theory. Loop quantum gravity advances a model in which, among other things, space is quantized, digital or granular in character and, therefore, is hypothesized to consist of a network of woven loops on the level of the Planck length ($10^{-35}$ meters) that are finite in character and, over time, form structures that are described in terms of "spin foam".

For example, numerical simulations have been done involving loop gravity in which space, itself, becomes a powerful force of resistance to gravitational collapse. According to such simulations, a quarter of the star was ejected due to the dynamic between gravity and space, and when this occurred, various kinds of radiation and particles (high energy gamma rays, cosmic rays, and neutrinos) were released.

The energies, types, quantities, and so on of such radiation/particles depend on the quantum theory being considered. Consequently, if naked singularities actually existed, one might have a basis for experimentally differentiating among such theories.

The theory of naked singularities is – at this point – highly speculative. It is unknown whether, or not, they exist, or if they do, whether, or not, they will be able to emit any sort of radiation or particles that can be detected and, then, be used as a way to distinguish between viable and problematic theoretical accounts of gravitational collapse.

Up until fairly recently, the event horizon of a black hole was considered by many individuals to be a rather innocuous entity. In other words, various physicists understood the event horizon as merely marking a point of no return ... an invisible, dark boundary that, once it was traversed, trapped particles and radiation – irrespective of the energy and velocity associated with such particles and radiation – permitted them to be drawn inexorably inward toward the gravitational singularity that existed at the core of a black hole.

According to the foregoing physicists, if a person were to fall through the event horizon, that individual would not observe anything of a strange nature. Such individuals could run experiments (until oblivion struck) that indicated the laws of physics operated the same on the inside of the event horizon as they did on the outside of that boundary surface.

Relatively recently, however, some physicists have challenged the idea that nothing much happens at the event horizon, They believe that a fierce dynamic takes place along the boundary surface that marks the outer reaches of a singularity.

Even before the foregoing sort of theoretical troublemakers came along, there was a controversy concerning black holes that seemed to indicate that either quantum physics or general relativity was wrong in their respective depictions concerning the nature of reality. The new kids on the block are suggesting that the issue might not be a matter of either one theory or the other being in error, but, perhaps, the problem is that both theories might be lacking in certain respects.

Let's begin at the beginning ... sort of. In 1974, Stephen Hawking pointed out a possible problem in relation to the standard model of black holes.

According to quantum theory, virtual particles (consisting of various particles and their antiparticle counterparts) are constantly popping into and out of existence. Hawking asks us to imagine a case in which such pairs show up just outside of the event horizon for a given black hole.

Hawking asks us to further suppose that one member of the virtual pair might fall through the event horizon, while the other member of the pair escapes being sucked past the invisible boundary and toward the core of the black hole. In the process of escaping, one particle carries off some of the energy/mass of the black hole, and this process is known as "Hawking evaporation" or "Hawking radiation".

Under normal circumstances, Hawking evaporation is something of a moot point. There is much more mass that is sucked into a black hole than evaporates from it.

However, Hawking next considers what might happen in the case of an isolated black hole. Given enough time, presumably, evaporation would take away more mass than was entering the black hole.

Hawking believes that such a thought experiment reveals the presence a problem. Although Einstein didn't think the conditions necessary for a black hole to form would ever be able to arise because there was too much instability associated with such a process, many theorists believe that the general theory of relativity indicates that the geometry of black holes will be smooth, and, as a result, every black hole of a given mass will have exactly the same kind of spin and charge.

However, Hawking claims the radiation that is giving rise to evaporation seems to indicate that the black hole has a temperature. Generally speaking, temperature is due to the motion of particles within a given volume that is enclosed or contained.

According to Hawking, the fact that a black hole appears to have a temperature suggests that a black hole consists of some sort of microstructure or set of discrete building blocks. Moreover, such discrete entities can be represented in terms of bits of information.

Hawking, and independently, Jacob Bekenstein, developed a formula for calculating the number of bits of information in a black hole. This formula is said to measure the entropy of a black hole.

According to some individuals, entropy is an index of the disorder that is present in a given system. As the possible states associated with such a system grow, then, for individuals who think along the foregoing lines, the entropy or disorder inherent in that system grows as well.

General relativity seems to suggest – at least for some individuals -- that black holes give expression to a smooth geometry, and therefore, cannot exhibit the property of entropy. Quantum mechanics, on the other hand, appears to indicate that black holes have a structure that gives expression to a degree of entropy (possible modes of arrangement) that can be measured.

Raising, and then reflecting upon, a few questions might be in order at this point. For example, do the events taking place at an event horizon really constitute something that is occurring within a black hole, and do those events necessarily imply that a black hole has some sort of microstructure rather than just indicating that the event horizon portion of the black hole might have some sort of structure?

An event horizon might be a very complex structure. How does one differentiate between what is, and is not, part of that boundary surface?

Can one suppose that singularity pulls with an equal force in relation to all aspects of an event horizon? Some numerical simulations have indicated that singularities do not always form in the geometric center of a given volume, and, therefore, the force exerted along the event horizon might not always be the same, and, as well, one might also have to take into consideration the possibility that events transpiring within the singularity or within the volume between the core singularity and its event horizon might attenuate the force of gravity, from place to place, as one moves along the event horizon boundary.

In light of the foregoing considerations, is it possible that upon contact with the outer portion of an event horizon, a particle still might be able to escape being drawn inward? On the other hand, that same particle (or one like it) might not be able to escape if it either entered deeper into the boundary complex of an event horizon or made contact with a point along the event horizon boundary that involved maximum, gravitational attraction.

Are the possible arrangements that might occur at the event horizon an indication of the degree of entropy or disorder present at that event horizon? Whatever the number of such possible arrangements might be isn't the basic issue a function of what actually takes place rather than a function of what might take place?

A wave function indicates that any number of events might be possible in a given context (say at some point along the event horizon). Yet, only one thing actually happens at such a point.

Therefore, the methodological prediction that many arrangements are possible in such a context doesn't necessarily accurately reflect the ontological nature of the situation except to the extent that the wave function can lead to the generation of a value that might turn out to be correct. This is sort of like a judicial system in which many possible charges might be thrown at a suspect, but, in the end, oftentimes, only one of those charges goes to trial or leads to conviction.

The process of methodologically filtering possibilities in order to be able to arrive at what, ontologically speaking, actually occurs (such as with a wave function or in conjunction with the charging process in a judicial system) does not give expression to the presence of disorder. Increasing the number of possibilities that might occur does not somehow make what does happen more disordered.

If one does not know how something ultimately works (such as in quantum mechanics), increasing the number of possibilities is more of an indication about the disordered state of one's understanding concerning the actual nature of a given context than it is an indication of the disordered state of that context. The nature of reality is ordered (after all, how does one explain the presence of physical laws if this were not the case), but our understanding (or underlying methodology) does not necessarily reflect the presence of that order.

Hawking assumes -- as do most astrophysicists – that virtual particle activity takes place at the event horizon. The amount of virtual activity that is predicted to occur in a vacuum is many orders of magnitude greater than what has been determined to be present experimentally, so the question really is: To what extent does virtual activity – to the extent it takes place at all – actually occur at any given point along the event horizon?

Notwithstanding the foregoing question, if one were to assume that virtual particle activity is occurring all along the event horizon, then why only consider cases in which virtual particles take mass/energy away from the event horizon of a black hole? Why not consider instances in which black holes take away mass/energy from the sea of virtual particles that supposedly pop into and out of existence just outside the event horizon ... such as the virtual particle cited in Hawking's aforementioned thought experiment that fell into the black hole?

Given a sufficiently complex event horizon boundary, it is possible there might some sort of dynamic equilibrium of virtual particle flow (to whatever extent it exists) that arises in and around an event horizon. Given a sufficiently complex event horizon boundary, one cannot automatically assume that the flow of virtual particles is only capable of traveling in one direction in relation to such an event horizon (that is, in the direction of evaporation away from the event horizon).

Even if there is no dynamic equilibrium present in and around an event horizon boundary, the rate of evaporation will be affected by the amount of virtual particles that actually do pop in and out of existence Furthermore, that rate of evaporation or radiation will also be affected by the number of those virtual particles that actually do end up taking away mass/energy from an event horizon.

Hawking believes that if a black hole existed in isolation and if, in addition, enough time were to pass, then, eventually, a black hole would lose all of its mass/energy to such an evaporation/radiation process. However, what if there were virtual particle related events (or even non-virtual particle related events) taking place inside a black hole that countered the effect of such evaporation/radiation?

The slower the rate of evaporation/radiation, the more plausible the possibility becomes that there might be offsetting virtual particle events (or non-virtual particle events) within that black hole. However, too little is known about: (1) The nature of singularities; (2) event horizons; (3) the dynamics within the volume between singularities and their event horizons, and, finally, (4) the collective interactional dynamics of (1) – (3) to be able to fix a rate of evaporation or radiation with any degree of certainty.

In addition, the idea that mass/energy might be taken away from a black hole by a virtual particle presents something of a problem. Supposedly, nothing can escape from a black hole, so how does some of its mass get radiated away?

The only way in which the idea of Hawking radiation or evaporation appears to make sense is if the radiated mass/energy comes from the dynamics of the event horizon rather than from the black hole per se. In other words, if the structural character of an event horizon boundary were sufficiently complex, then, such complexity could entail the possibility that, from time to time, a certain amount of radiation or evaporation might be taken away from just the event horizon without actually reducing the mass/energy of the singularity itself.

If the foregoing scenario were correct, then the role of Hawking radiation or evaporation operates in conjunction with the dynamics of the event horizon boundary structure and is not necessarily a function of a black hole considered as a whole. The transfer of heat that occurs at the event horizon does not constitute the temperature of the black hole, but, instead, gives expression to the temperature of the event horizon.

Calculations indicate that the magnitude of the temperature associated with event horizon dynamics is in the order of billionths of a degree Kelvin for a stellar mass that has collapsed and formed a black hole containing such an event horizon. Detecting the presence of such a temperature would be extremely difficult, if not impossible, to achieve.

Temperature is a measure of the radiation that is given off through the dynamics of interacting particles, irrespective of whether, or not, such interaction takes place inside something ... such as a black hole. There is heat being generated through the dynamics that occur as particles interact with the event horizon (which is not necessarily an inert, empty boundary surface), and those particles either pass through that boundary or ricochet off it ... that is, energy is being transferred (e.g., in the form of, among other things, heat) during the course of such transactions.

What takes place at an event horizon does not necessarily imply that the interior of a black hole is structured ... although it might be. What takes place at the event horizon reflects the structure of the event horizon in terms of whatever graviton particles, virtual particles, and in-falling particles that are present at that boundary interact with one another.

What takes place at the event horizon is not a disordered process. Rather, quantum and gravitational forces/particles come together and interact in determinate ways, irrespective of whether, or not, we are able to predict what the nature of that interaction will be.

Interpreting entropy as a measure of disorder that is marked off in bits of possible arrangements is to impose something on the dynamics of the event horizon that do not necessarily reflect the kind of order that is present in the underlying dynamics. For instance, when virtual particles are separated at the event horizon, and one particle falls into the black hole while the other particle escapes, what is the degree of entanglement between those two particles?

Is entanglement something with which a black hole interferes? To whatever extent entanglement exists despite separation by an event horizon, then, to that extent, disorder is not present, and to describe such a situation in terms of entropy seems problematic and misleading.

Entropy is an indication of the presence of irreversible events. While irreversibility restricts the kinds of order that are possible as one moves forward in time, I am not sure that such irreversibility counts as being an expression of disorder, but, in fact, such a condition seems more like a statement concerning the nature of the order that is manifested in such a system as it unfolds across time.

Entropy might shift the ratio of what is still possible versus what is no longer possible (barring the intervention of external forces). Nonetheless, this is not an increase of disorder but a shift in the way that existing order manifests itself under a given set of conditions.

If one uses the Hawking/Bekenstein definition of entropy, the more choices one has – that is, the more ways that current possibilities might be arranged, then, seemingly the more disordered such a situation is. And, yet, choice is the ordered activity (even if done poorly) that breaks the foregoing symmetry of possibility by selecting (as a function of rules, principles, or certain valuations) for one of those possibilities ... indicating that disorder is a somewhat illusory way of looking at such a situation.

Choice has consequences that often are irreversible. In this sense, entropy does increase as a function of the kind of order that choice carves out of existing possibilities despite the presence of whatever irreversible features might have entered the picture as a result of those choices.

Presumably, particles don't make choices. However, their dynamics do give expression to events that become ontologically or existentially irreversible (even if, mathematically speaking, the laws of physics say that, in principle, those events are reversible).

The dynamics of particles, whatever they might be, give expression to order, not disorder. Such dynamics do generate entropy

| Cosmological Frontiers |

161

because they involve events that, subsequently, cannot be reversed without outside intervention (and, sometimes, not even then ... as is the case, for example, in the case of particles that traverse the event horizon of a black hole).

However, the foregoing entropy is a reflection of the changes in the ratio of what is still possible versus what is no longer possible. This gives expression to the presence of a certain kind of restricted order rather that a state of disorder.

As soon as one makes entropy a measure of the number of arrangements that are possible for a given set of events, one makes entropy the function of a system of measurement. Such a process might reveal more about the properties of that sort of methodological system than it provides insight into the dynamics of that which is being measured.

Conditions of irreversibility do occur, and, therefore, when considered in terms of irreversibility, the notion of entropy does offer some insight into what is transpiring. However, when entropy is filtered through a measurement system rooted in information theory, then, the process seems rather arbitrary, artificial and forced.

What is the relationship between a singularity and its event horizon? We don't know because what goes on within a singularity or the event horizon is unknown.

If singularities generate some sort of emergent phenomena, then, there is a possibility that such processes might be able to counteract the effects of gravity in certain ways. If this is the case, then, the structural properties of the event horizon could reflect those sorts of anomalous, internal dynamics.

Furthermore, given that the volume through which the Schwarzchild radius runs is filled with moving, charged particles, one might predict that such a volume could give rise to plasma phenomena, and, therefore – possibly -- powerful, electromagnetic currents. Such phenomena, if they occur, could affect what happens at the event horizon.

One can't really talk about Hawking evaporation or Hawking radiation independently of what is, or isn't, taking place in, and around, singularities. And, since no one knows what is transpiring in singularities or within the volume through which the Schwarzchild radius runs, or in conjunction with the event horizon for such a singularity, any statements about what Hawking evaporation or Hawking radiation will, or won't, do – even in isolated contexts -- seems rather premature.

According to Hawking, the structural properties of black hole-evaporation or black hole-radiation is not a function of the material that makes up the black hole from which it supposedly escapes. Since nothing can escape from a black hole – and, presumably, this includes information – then, from Hawking's perspective, the information contained in such evaporation or radiation is unrelated to the information that is contained in a black hole.

Information is a way of describing a system. Various methods for measuring information will describe such a system in different ways.

Information is not the same thing as that which is being described by means of the sort of theory of measurement that is inherent in a given notion of information. Therefore, when Hawking speaks about the information contained in a black hole or about the information contained in Hawking evaporation or radiation, he is talking about a form of measurement and understanding that does not necessarily accurately reflect the nature of what is being measured (although that form of measurement might be heuristically and/or descriptively valuable in various ways).

Information does not escape from a black hole because information was never in that black hole. Information was always external to such an entity in the form of a set of methodological processes through which to analytically parse the nature of the contents of such a black hole.

Hawking evaporation or radiation is not made up of information. Rather, information is a way (but not necessarily the only way) of describing that sort of evaporation or radiation.

Information does not become lost in a black hole. What goes missing in action is out ability to access the interior of the black hole so that one can try to generate descriptions (perhaps through some kind of information theory) concerning what the ontological properties of such a hole are.

Furthermore, if what I have said previously about the dynamics of the event horizon is correct – namely, that Hawking evaporation or radiation is a function of an event horizon and is not a function of the associated black hole per se – then, informational descriptions of such dynamics never entail anything being lost. Moreover, Hawking is not necessarily correct when he claims that the informational properties used to describe such evaporation or radiation are unrelated to what is transpiring within a black hole since the nature of Hawking radiation or evaporation might be dependent on the way the internal dynamics of a black hole affect the structural properties of its event horizon and, therefore, how, when, to what extent, and in what forms Hawking radiation or evaporation might arise.

Hawking believes the issue of information leads to a paradox. Either one has to modify quantum theory in order to allow for the loss of information, or one has to modify relativity theory to account for, among other things, the entropic features (which are rooted in information theory) that are associated with the temperature of a black hole and, therefore, suggest that black holes give expression to something other than the smooth geometry that is proposed by general relativity.

I'm not convinced that one is forced to choose between the foregoing two options. The manner in which Hawking interprets the idea of information is clouding the issue ... that is, Hawking seems to reify information and make it essential to the ontological structure of something, when, information is only a methodological construct that is being used to analytically frame and filter reality.

Juan Maldacena put forth an idea in 1997 that sought to resolve Hawking's alleged paradox. The idea is rooted in what is known as the Maldacena duality.

A duality exists when two things that seem to be quite dissimilar can be shown to be equivalent to one another. Maldacena purported to show that the mathematics of a theory of quantum gravity based on string theory is equivalent to the mathematics of ordinary quantum mechanics under certain conditions ... such as those existing in conjunction with a black hole.

If the foregoing perspective is true, then, the laws of quantum mechanics can be applied to general relativity. This means, in turn, that if information cannot be lost in quantum mechanics, then, information cannot be lost in conjunction with gravitational systems.

One implication of the foregoing is that Hawking was wrong when he claimed that information is lost because Hawking evaporation or radiation is independent of the make-up of a black hole. Apparently, in some fashion, information contained in the black hole escapes with such evaporation or radiation.

Again, I believe the problems surrounding the alleged information paradox are ill-conceived. Treating information as if it were an ontological "stuff" rather than the methodological concept it is, introduces unnecessary difficulties into trying to determine the nature of the relationship among black holes, quantum mechanics, and general relativity.

As previously indicated, Information can't be lost in a quantum system because the latter doesn't contain information. Rather, a quantum system has structural properties and dynamic forces that can be described, well or poorly, through various frameworks of information theory.

Information doesn't evaporate or radiate from a black hole. Information is a methodological means that can be used to describe some of the properties of such evaporation or radiation.

The Maldacena duality also indicates that space-time will be a function of a different set of processes than we perceive to be the case. According to Maldacena, space-time is a three-dimensional, holographic projection of a more fundamental two-dimensional surface of a sphere.

How physical laws become encoded in such a two-dimensional surface is unknown. How the encoded order of a two-dimensional surface becomes projected as a three-dimensional image is also unknown.

Furthermore, as noted in the last chapter, space-time does not give expression to ontology. Instead, space-time constitutes a mathematical way of giving representation to the events of curvature occurring within a gravitational field.

Approximately 20 years ago, Leonard Susskind and Gerard 't Hooft proposed their own solution to the Hawking information problem. Their proposal became known as "black hole complementarity".

According to the two physicists, no information is ever lost in conjunction with black holes. If someone were to pass through the event horizon of a black hole, they would see whatever information came into the system, whereas, external observers would observe information come out of that same system through the process of Hawking radiation or evaporation.

There is a parallel between the way in which space-time is understood (or misunderstood) in general relativity (e.g., Maldacena's remarks in the previously discussed Maldacena duality) and the manner in which information is interpreted in conjunction with black holes. Both are artifacts of an underlying methodological framework that is seeking to transform a methodological concept into some sort of concrete, ontological reality, and, in the process, understanding becomes confused and error-ridden.

Notwithstanding the foregoing considerations, between the work of, on the one hand, Susskind and 't Hooft, along with, on the other hand, the efforts of Juan Maldacena, the so-called information paradox problem originally posed by Hawking in the 1970s is considered by many physicists – including Hawking himself – to have been laid to rest. However a new variation on that problem has arisen.

This latest difficulty involves the issue of entanglement. According to quantum theory, entanglement can only involve two particles.

If particle 'A' is entangled with particle 'B', then, neither 'A' nor 'B' can be entangled with any other particle. The foregoing belief is not the result of empirical demonstration, but, instead, reflects the presence of an underlying assumption.

However, for the sake of argument, let's treat that assumption as if it were true. What happens to the pair of virtual particles in Hawking's original thought experiment involving two particles that arose together and then become separated ... with one of the two particles disappearing beneath the event horizon, while the other carries away mass from the associated black hole in the form of Hawking evaporation or radiation.

Some physicists claim that the two particles are entangled. Since we don't really understand all of the ins and outs of entanglement, one cannot necessarily be certain of the conditions through which entanglement does – and possibly – doesn't occur.

However, let's assume that the two particles are entangled. Will the two particles continue to remain entangled?

Initially, Hawking was talking about virtual particles that arise near the event horizon of a black hole. Virtual particles pop into and out of existence.

The virtual particle that popped into existence and, then, disappeared into the black hole does not necessarily continue to exist since virtual particles are short lived. Once the virtual particle inside the black hole pops back out of existence, is it still entangled with the virtual particle that carries away mass/energy from the event horizon?

We don't know. We don't even understand why and/or how virtual particles pop into and out of existence ... if that is what actually happens.

Quantum physics predicts the existence of such particles. Nonetheless, quantum physics doesn't explain the dynamics underlying those processes of popping into and out of existence.

Let's assume that such entanglement does continue on in some unknown fashion. What then?

Is it possible for a particle to be entangled simultaneously with more than one particle? Quantum theory claims this is not possible, but it provides no evidence to back up such a claim.

Conceivably, at any given time, only two particles might be able to become entangled (although we don't know this for sure). However, even if the foregoing restriction applies, does anything prevent such particles from subsequently becoming entangled with other particles?

Maybe, particles have some sort of quantum memory concerning what particles they have engaged. Perhaps, there are limits to what can be contained in such quantum memory, but, there doesn't seem to be any a prior reason why such a memory – if it existed in some form – couldn't contain a history of more than one such engagement.

The phenomenon of entanglement has been demonstrated to occur. Consequently, there must be something like quantum memory that links the two particles together and renders them sensitive to what is transpiring in one another.

Information theory might be used to describe such entanglement. What is essential, however, is the ontological character of that entanglement and not the way in which it is described.

Entanglement keeps track of what its nature permits. The particles that are entangled will continue to be sensitive to the presence of that condition.

Which particles become entangled with one another is unknown to us. In general terms, however, one could say that whatever particles become entangled in relation to the interior of a black hole -- along with the particles involved in the Hawking radiation that are emanating from the event horizon associated with such a black hole – then, we do not have any viable reason to suppose that such particles will not continue to maintain their relationship of entanglement.

One should keep in mind that entanglement is based upon statistical arguments. Entanglement alignments in experiments indicate that what is observed is greater than a certain value but do not indicate that all such alignments give expression to entanglement.

If particles that are entangled could have more than one such entanglement relationship, there might be instances in which an entangled particle gives preference to one kind of entanglement rather than another. In fact, maybe one of the reasons why entanglement alignments do not show up as being 100% in experiments is because of such multiple entanglements and the fact that only one entanglement relationship at a time can be expressed.

Conceivably, over time, various entanglements might form, and, then, on any given occasion, one kind of entanglement, rather than another, might be activated. Or, possibly, entanglements might be serial in nature, with subsequent entanglements taking the place of previous entanglements (sort of like a process of serial monogamy).

One cannot say whether, or not, the forces associated with a black hole are capable of breaking the bonds of entanglement. Moreover, if such forces are capable of doing this, we do not understand the nature of the dynamics associated with such a process of entanglement disruption.

According to some theorists (for example, Joel Polchinski), the cost of breaking quantum entanglements would require energy. If quantum entanglements were capable of being broken, such theorists indicate that this would imply that the event horizon might consist of a ring of fire involving high-energy particles that were responsible for the breaking of entanglement bonds.

As noted earlier, whether, or not, the bonds of entanglement can be broken (via energy or in some other way) is unknown. Furthermore, even if entanglement relationships can be broken, there is nothing currently known that requires such events to take place along the event horizon (as opposed to taking place through some other dimensional dynamic), and, as well, there is nothing that is currently known which requires that changes in entanglement orientation must take place by means of high-energy particles (rather than through some other set of, say, thermodynamic conditions – such as an unstable decay process -- that is conducive to the dissolution of entanglement bonds).

Consequently, proposing the existence of a firewall along an event horizon does not necessarily follow from the idea that entanglement bonds can be broken or decay under certain conditions. Nonetheless, at the very least, all of the foregoing possibilities indicate that the event horizon might be a very "happening" environment … even if it does not necessarily constitute a high-energy firewall of sorts.

Polchinski does indicate that if such firewalls do exist, they will be very hard to detect. He believes that the gravitational attraction of the singularity would tend to cloak, to some extent, the presence of such high-energy events and, thereby, render them less visible to external observers.

None of the foregoing considerations indicate whether, or not, black holes actually exist. Those considerations give expression to interesting possibilities, but they all are predicated on the assumption that black holes are real.

In early 1974, two researchers – Robert Brown and Bruce Balick – detected the presence of a very compact source of radio waves that was emanating from a point near the center of the Milky Way galaxy, some 26,000 light years from Earth. They named the source: Sagittarius A* (A-star).

Upon further study, the area in the vicinity of Sagittarius A* was found to contain groups of stars with anomalous motions. In addition, there streams of agitated gas were present in the vicinity of Sagittarius A*, and, X-rays were detected in that area as well.

By tracking the aforementioned anomalous motions of stars near Sagittarius A*, researchers were able to construct a picture of the object that appeared to be responsible for such anomalous motions. The mysterious cosmic object was calculated to have a diameter of approximately 15.5 million miles, with a mass that was 4.3 million times that of the Sun.

Sagittarius A* is considered to be a black hole. Yet, relatively recently, data has been collected indicating that there might have been times in the past when Sagittarius A* was much more energetic than presently appears to be the case.

When black holes encounter galactic materials, those materials begin to swirl about the event horizon of the black hole and, at a certain point, form what are known as "accretion disks". As those disks are drawn toward a black hole, they become heated and, as a result, radiate intensely.

At the present time, Sagittarius A* is fairly dim. This suggests that there is relatively little galactic material to be found near that body.

However, a French astrophysicist – Maïca Clavel – has been conducting research on Sagittarius A* for more than twelve years (1999 – 2011). She interprets the data she has compiled as indicating that the black hole has been much more active in the past than presently is the case.

For example, Clavel has been investigating X-ray data associated with Sagittarius A*. Clavel refers to the series of X-rays she has been exploring as "light echoes" – that is, X-rays which were emitted at some point in the past when there was intense activity in the vicinity of Sagittarius A* (e.g., an accretion disk of galactic material that was being drawn in toward the black hole).

According to Clavel, the aforementioned X-rays were emitted and, then, they proceeded to interact with iron atoms in some of the surrounding gas clouds, and, then, eventually, made their way to Earth. Those X-rays constitute windows into the past of the area in which Sagittarius A* is located.

If one analyzes the patterns of X-rays emanating from the Sagittarius A* region, as Clavel has done, one discovers that the pattern of light echoes involving such X-rays has changed from time to time. Clavel believes that the data she has analyzed shows there have been several periods of intense activity involving Sagittarius A* that have occurred over the last two to three hundred years.

Such periods of intense activity are considered to give expression to instances in which Sagittarius A* came in contact with stars, gasses, and/or planets. After those materials were completely consumed by the black hole, the cosmic giant entered into a period of relative quiescence similar to what appears to be the case at the present time.

However, for the last year, or so, a giant gas cloud – designated G2 – has been speeding by Sagittarius A* at a velocity of approximately 6.2 million miles per hour. Since, the nearest point of approach of G2 in relation to that black hole is estimated to be roughly 15.5 billion miles distant from the event horizon of Sagittarius A*, G2 will not be consumed by the putative black hole.

Nonetheless, Sagittarius A* and G2 are interacting with one another. The gas cloud has been stretched and fashioned into a filament, of sorts, that wraps around – at a distance – the black hole.

Some theorists believe there are a number of much smaller black holes that inhabit the general region in which Sagittarius A* is found. These black holes are not considered to be much bigger than a city (with masses roughly a few times that of the Sun), but they are thought to be fairly numerous (20,000 or more of them).

Although tied, to some degree, to actual empirical data, interpretive estimates involving the number of smaller black holes surrounding Sagittarius A* are fairly speculative. Nonetheless, if such mini-black holes do exist, theorists believe those cosmic entities might interact with the G2 gas cloud and, in the process, steal some of the material from that cloud.

Such thefts likely would give rise to the generation of X-rays. In turn, those emanations might be detectable through instruments like the Chandra X-ray Observatory.

In addition, a project is underway to link together a dozen observatories around the world. The network of observatories is referred to as the Event Horizon Telescope, and the intent underlying the forming of such a network is to be able – hopefully – to obtain a closer, more detailed look at the event horizon of Sagittarius A*.

Over the last 40 years, the idea of black holes has crept into many corners of consciousness, both professionally and publically. There are many reasons for this.

The notion of a black hole – which is present in Schwarzchild's very first solution to Einstein's theory of general relativity – is inherently intriguing. Do they exist, and if they do exist, what are their properties? These are questions that tug at the innate curiosity of human beings – both professional and otherwise.

However, there is another dimension to the issue of black holes that bears upon the question of origins. If one -- as Lemaître did more than 80 years ago – were to rewind the physics of an expanding universe (and the universe might, or might not, be expanding), that rewinding process supposedly brings us back to a starting point in which the entire universe is compressed into an infinitesimally small point known, currently, as a singularity.

The foregoing singularity has all of the characteristics of being the mother of all black holes. Or, perhaps, more accurately, it might have been the mother of all singularities since it is uncertain whether, or not, an event horizon would form in conjunction with such a singularity (e.g., Did space exist outside of such a singularity?).

Since that singularity is, allegedly, infinitely small and infinitely dense, we have no idea what the dynamics, if any, of such an entity might be. We do not have the physics to describe that singularity ... our physics breaks down under those circumstances.

What happens, if anything, to space and time under such conditions? We don't know!

What happens to leptons, quarks, and the carriers of force when subjected to infinite amounts of gravitational attraction? We don't know!

Is there some limit to how densely packed baryonic materials can be in a singularity? We don't know?

Was gravitational attraction necessarily the dominant force in the primal singularity? We don't know!

Can one assume that all of the matter and energy in the universe is capable of being compressed within a point singularity? We don't know!

Were there emergent properties that existed in conjunction with the primal singularity? We don't know!

Is gravity holding such a singularity together, or is some other emergent phenomenon holding things together ... or, did some combination of the two (i.e., emergent properties and gravitational attraction) keep the primal singularity intact? We don't know!

What would induce the foregoing sort of singularity to explode or break apart? We don't know!

If the original singularity exploded in a Big Bang, does this have any implications for what the fate of "conventional" black holes" might be? We don't know!

Was the primal singularity stable or unstable? We don't know!

Are the elements present in that singularity – whatever those elements might be – entangled, and, if so, in what way are they entangled? We don't know!

What effect, if any, would entanglement have in shaping a Big Bang? We don't know!

Could the original singularity provide a means through which the universe, as we experience it, is the result of a holographic projection of some kind? We don't know, and even if it did, we don't have any idea how such a holographic process was initially encoded or projected.

Does such a singularity contain the sorts of symmetry-breaking elements that would be capable of inducing some sort of unified force to proceed, at certain junctures, along divergent pathways and, thereby, give birth to the four forces with which we are familiar (electromagnetic, strong nuclear force, gravitational force, and the weak force that is present in such phenomena as radioactive decay)? We don't know!

If there were some sort of unified force existing in conjunction with the primal singularity, would that unified force necessarily have given expression to gravitational effects, or would gravitational effects have arisen through a currently unknown mode of symmetry breaking at some point during the Big Bang? We don't know!

Does the dissolution of the singularity turn on the Higgs field? We don't know!

Does the dissolution of the singularity turn inflation on and off? We don't know, and even if it did, we have no idea how such an on and off switch works.

None of the theoretical work that has been carried out in conjunction with the nature of "conventional" black holes – assuming that they exist – is able to shed any light on the nature of the dynamics that might exist in the primal singularity that allegedly existed prior to the issuing forth of the universe by means of a Big Bang. Not only is the work of astrophysicists in relation to black holes often highly speculative but, as well, the possible properties that have been determined seem rather superficial in nature and appear to involve, for the most part, just the dynamics associated with the event horizon of a black hole rather than anything of substance concerning the dynamics that might be present in the singularity that occupies the core of a volume for which the event horizon serves as an outer surface boundary.

Many proponents of the Big Bang assume that the temperature of the primal singularity was astronomical. However, much – if not all -- of the data on which such an assumption is based is fairly circumstantial and, more often than not, seems to be filtered through a hermeneutical framework that needs to conceptually build certain capabilities into the primal singularity in order for a given theory of the Big Bang to appear to have some degree of plausibility.

Many proponents of the Big Bang believe that whatever symmetry breaking events occurred in relation to that event were, in some sense, energy dependent. They further believe that as the energy of the universe dropped, different forces precipitated out of the original unified force field that existed, but, to date, no one has worked out what the precise nature of such a process of symmetry-breaking would involve and whether, or not, that process had anything of an essential nature to do with the issue of temperature (i.e., whether temperature was causal or merely an unconnected correlation).

Steven Weinberg's book, *The First Three Minutes*, provides an intriguing and technically detailed narrative for how things might have transpired following a Big Bang (assuming the latter event actually took place). Nonetheless, there are too many unknowns surrounding the set of initial conditions involving a primal singularity that allegedly existed prior to a Big Bang (many of which have been touched upon in the preceding two to three pages) to be persuaded that Weinberg (or other Big Bang advocates) has accurately captured the highlights of such a process of cosmic unfolding.

None of the foregoing is meant: (1) to suggest there was no Big Bang; or, (2) to dismiss, out-of-hand, the possibility that Steven Weinberg might have been right in part, or wholly, with respect to his account of the first three minutes of, or following, a Big Bang involving a primal singularity that supposedly took place at some point in cosmic history. Rather, what is being stated is the following: If there were such a primal singularity with an ensuing Big Bang, we have almost no understanding – and, perhaps, no real understanding at all – with respect to how things might have unfolded in conjunction with an explosive event of some kind involving such a singularity.

Chapter 7: Through A Glass Darkly

In 'Chapter Five: The Electric Universe', a perspective was outlined indicating there were some scientists who believe there might be ways of explaining the anomalous movements of stars within galaxies as well as galaxies in relation to other galaxies without necessarily having to resort to the notion of dark matter in order to explain those sorts of behaviors. The time has come to provide some of the structural features of a dark matter framework, and, thereby, offer a partial overview of what the cosmological terrain looks like when viewed through the filters of dark matter.

Dark matter, if it exists, is largely invisible to us. It doesn't seem to interact with those portions of the universe that are described through the Standard Model of quantum physics. Therefore, with the exception of the gravitational impact associated with dark matter, currently, there seems to be no way to detect its presence.

Moreover, dark matter need not necessarily be restricted to the distant horizons of outer space that exist within, and beyond, the Milky Way galaxy. In fact, depending on what dark matter turns out to be (if it exists), it could be present – to some degree – in the spaces where one eats, sleeps, works and plays, and, yet, we might remain unaware of its presence.

The astronomer, Fred Zwicky, appears to have been the first individual to propose the idea that something like dark matter might exist. Based on the observational data associated with the rotational velocities of, among other things, galaxy clusters, Zwicky came to the conclusion in 1933 that introducing some form of additional source of mass might be necessary in order to be able to explain the foregoing kind of rotational behavior … something that could not be accomplished if one were to take into consideration only the mass of the visible stars in those galactic clusters.

Galaxies were moving faster than they should have if those galaxies contained just the mass that was present in their visible stars. Therefore, perhaps some form of dark matter might account for the mass that seemed to be missing and that, if present, would make sense of the rotational velocities that were being observed.

Seventeen years later, Vera Rubin, in her master's thesis, came to a conclusion similar to that of Zwicky. Using a specific frequency of light that is emitted by hydrogen atoms, Rubin analyzed the motions of a variety of stars and, then, used the data associated with the motion of the individual stars to construct a picture of the movements of the galaxy or cluster of galaxies containing such stars.

Her data and calculations led her to advance the idea that there might be something present in the rotational motions of galaxies and galaxy clusters that was affecting their motions. This 'something' seemed to encompass an enormous amount of mass ... indeed, more mass than was contained in all of the stars present in the galaxies and galactic clusters being studied.

Her master's thesis was accepted, but her ideas were largely ignored. She went through a similar sort of experience in conjunction with her doctoral thesis.

In order to survive professionally, Rubin pursued more mainstream topics of research. However, twenty years after completing her master's thesis, she returned to the issue of dark matter.

In 1970, Rubin, together with W. Kent Ford, sent an article to the *Astrophysical Journal*, and, eventually, the article was published. The paper focused on some of the rotational properties of the Andromeda nebula.

According to Rubin and Kent, in order for the stars of the foregoing nebula to be moving with the high velocities being observed, there would have to be up to ten times the amount of mass in that galaxy than could be accounted for by the stellar objects that were visible. Apparently, 90% of the matter in the Andromeda nebula consisted of dark matter.

For more than half a century, astronomers have been aware that the outer portions of the Milky Way galaxy (that is, between 50,000 and 75,000 light years from the center) are warped (by as much as 7,500 light years out of the galactic plane), exhibiting undulations like an old-fashioned vinyl record that had been put in a fire for a period of time. Although many astronomers attributed the foregoing undulations to the gravitational influence of the Large and Small

Magellanic Clouds, fully accounting for the structural properties of those undulations was an on-going problem ... that is, until researchers began to factor in what impact dark matter might have on the shape of those outlying undulations.

However, while introducing dark matter into the galactic warping problem helped to make sense of the structural anomalies that existed in the outer regions of the Milky Way galaxy, invoking dark matter also created some problems. More specifically, when astronomers estimated the amount of dark matter that might be present in the far regions of the Milky Way galaxy and, then, fed those quantities into their computer simulation programs, the results indicated there should be hundreds, if not thousands, more satellite or dwarf galaxies than were being observed, and this became known as the missing satellites problem.

The discovery of galaxies known as ultra-faint dwarfs has reduced the extent of the problem a little. These galaxies often consist of just a few hundred stars, and therefore, tend to be detectable only through the use of specialized methods for analyzing astronomical data.

Such data-handling methods have filled in some of the blanks that were left over after noting the difference between what has been predicted and what has been observed with respect to the issue of satellite galaxies. However, notwithstanding the inclusion of the ultra-faint galaxy data, the discrepancy between the number of satellite galaxies that have been predicted and what actually has been observed is still considerable.

Dwarf galaxies have been uncovered out to a distance of 150,000 light years. Some individuals theorize that hundreds more of those kinds of galaxies might exist as far out as one million light years from the center of the Milky Way galaxy.

Empirical checks on the foregoing claims will not be possible until more powerful telescopes, like the Large Synoptic Survey Telescope in Chile, come on line. Unfortunately, the LSST is some 4-5 years away from becoming operational.

Furthermore, despite all of its technical advances, even when it comes on line, the LSST might not be able to resolve the foregoing issue concerning satellite galaxies. For instance, if some galaxies are made entirely of dark matter, then, sophisticated instruments like the LSST won't be of much assistance ... at least in any direct sense.

Astronomers, however, have demonstrated that, oftentimes, some of the largest undulations that are present in the outer regions of galaxies like the Milky Way are due to the influence of passing galaxies. There are techniques for distinguishing between the modes of perturbations that are caused by passing galaxies and the sorts of perturbations that might be generated through other sources.

Using the foregoing techniques, researchers are able to determine the location and mass of galaxies that are causing perturbations in the outer regions of a given galaxy. In fact, such techniques are sufficiently sensitive that influences from galaxies that are less than $1/1000^{th}$ the size of a primary galaxy can be detected.

Some researchers -- using the foregoing method -- believe they have detected the presence of a galaxy made from dark matter that exists some 300,000 light years from the center of the Milky Way. Steps are being taken to try to confirm the foregoing inference, but even if found, the existence of one dark matter galaxy might not be able to fully resolve the problem created by the difference between predictions and observations with respect to the dwarf satellite issue.

Based on a variety of considerations, dwarf galaxies appear to be more involved in the dynamics of dark matter than many other galaxies are. Why this should be the case – if it is – is unknown, but the search for dark matter galaxies could provide some insight concerning the nature of the relationship between satellite galaxies and dark matter.

Another issue involving satellite or dwarf galaxies has to do with their alignment. They often form straight lines.

Computer simulations based on traditional assumptions about how galaxies supposedly develop indicated that satellite galaxies should be uniformly distributed around their primary galaxies in a spherical-like formation. Yet, once again what was observed to be the case could not be reconciled with existing models of galactic dynamics.

More recent computer simulations – based on different assumptions and considerations -- paint a very different picture. The newer models indicate that galaxies are not randomly distributed in the universe but, instead, form a structure that is known as a "cosmic web".

This web consists of sheets made from millions of galaxies that stretch across hundreds of millions of light years. Filaments extend between the sheets.

Voids devoid of galaxies populate the extra-galactic volumes that separate filaments from one another. Large primary galaxies serve as hubs or anchor points for many of the filaments that run between the sheets of galaxies.

The aforementioned filaments have a higher density than the voids. As a result, they attract and draw together dust and gas and begin to form proto-galaxies.

The filaments appear to be regions containing higher concentrations of dark matter. The dark matter-laden filaments hold the dwarf, satellite galaxies in alignment during the process of galaxy formation, and at a certain stage of development, dwarf galaxies begin to be pulled toward whatever nearby galactic region is most massive, but do so in an aligned fashion ... as has been observed in conjunction with the way satellite galaxies appear to interact with the Milky Way galaxy.

If the foregoing account accurately reflects the structural character of the universe, it raises a lot of questions. For example, how did the filaments form and become connected to the galactic sheets in the way that seems to be the case, and why are the filaments distributed in such an ordered fashion?

Do the filaments get drawn into the galactic sheets through the gravitational pull of the hub galaxies? If not, then, why doesn't this take place? Is there more than one kind of gravitational force at work?

Do the filaments become incorporated into the areas of gas and dust accretion that collect at different points along the filaments and, subsequently, become dwarf or satellite galaxies? If not, why doesn't this happen?

Is it possible that similar systems of filaments – although on a smaller scale -- exist within galaxies and not just between galactic sheets? If so, is it possible that the theory of galactic formation proposed by Halton Arp (see: 'Chapter 2: The Meaning of Red') might

| Cosmological Frontiers |

involve an intra-galactic system of dark matter-laden filaments that link together various portions of a galaxy?

Wouldn't the movement of ions, gases and dust particles brought about through the pull of gravitational forces help give rise to plasmas and electromagnetic dynamics? Do the electromagnetic dynamics of plasmas figure into the formation of proto-galaxies and dwarf galaxies that takes place along the filaments of the cosmic web?

-----

Indirect evidence concerning the existence of dark matter (based on the presence of anomalous, gravitational phenomena) is one line of inquiry. Another line of inquiry is to try to determine the identity of dark matter.

Many research programs have been established to actively search for clues that might help elucidate the structural nature and properties of dark matter. Although many possibilities have been, and are being, considered, trying to pin down the essence of dark matter has proved, and is proving, to be a frustrating process.

One category of missing-mass candidates is known by the acronym: MACHO. This stands for: massive compact halo objects.

Among the possibilities that fall within this category are different kinds of cosmic objects. For instance, black holes, neutron stars, brown dwarfs, dead white dwarfs, and even extremely large, Jupiter-like planets are members of the MACHO category.

All of the foregoing entries have one thing in common. Despite their size and mass, they don't give off much, if any, light, and therefore, would be hard to detect.

However, the presence of a MACHO object might be detectable through the phenomenon of gravitational lensing ... a topic that was briefly explored in several previous chapters (see – 'Chapter Two: The Meaning of Red' and 'Chapter 5: Matters of Gravity'). Gravity has the capacity to affect light in various ways (e.g., bending it, focusing it), and, therefore, if one comes across a lensing event, one can try to determine whether, or not, that event was due to the presence of a MACHO member.

A project was set in motion to search for, among other things, possible gravitational lensing events. In 1993 astronomers at the

Mount Stromlo Observatory near Canberra, Australia began to look for instances of gravitational lensing.

If a suitable candidate were found, the Hubble Space Telescope would be used to try to determine what might be generating such a lensing phenomenon. In 2001, the Hubble Space Telescope identified a white dwarf some six hundred light years away as being the cause of a given instance of gravitational lensing.

Whether the white dwarf discovered in 2001 was the tip of a MACHO iceberg or merely an isolated event became something of a moot point. There is an issue – known as the nucleosynthesis problem – which poses a fundamental problem for those individuals who wish to argue that one might be able to account for dark matter through the MACHO category of cosmic objects.

The process of generating heavier elements from lighter elements is known as nucleosynthesis. In 1948, George Gamow, Ralph Alpher, and, possibly, Hans Bethe (there is some question as to whether Bethe actually participated) wrote a paper on how light elements such as hydrogen, deuterium, helium, and lithium might have formed following a Big Bang and made predictions concerning the relative abundance of the elements that were likely to be found.

Subsequently, empirical data accumulated indicating that the relative abundance of light elements observed in the universe seemed to confirm the predictions of Gamow, Alpher and Bethe. At the time, the apparent agreement between theoretical prediction and empirical results tended to lend considerable credibility to the idea of a Big Bang.

Later on, however, empirical determinations were made that indicated the abundance of several isotopes of lithium actually departed significantly from what had been predicted by Gamow, Alpher, and Bethe. This discrepancy is known as the 'primordial lithium problem' and tends to undermine the idea of a Big Bang ... at least as outlined by Gamow, Alpher and Bethe in their 1948 paper.

The lithium problem notwithstanding, the calculations made by Gamow and his colleagues in relation to the issue of nucleosynthesis entail a problem for MACHO-type accounts of dark matter. More specifically, if one were to consider all the matter in the universe that consists of protons and neutrons (the arrangements and combinations of such particles is at the heart of nucleosynthesis) that quantity of matter turns out to be significantly less than the amount of dark matter that has been calculated to exist.

Thus, even if many more gravitational lensing events were discovered that were caused by white dwarfs, or brown dwarfs, or black holes, or Jupiter-like planets, it wouldn't matter much. While some small percentage of dark matter might be a result of MACHO-related phenomena, the amount of dark matter in the universe appears to be significantly more than can be accounted for by not only MACHO-based accounts, but, as well, the amount of dark matter has been calculated to be greater than the amount of baryonic matter (involving various combinations of protons and neutrons) considered in its entirety.

If one eliminates the MACHO category from consideration, then, what other possibilities are there? Could entities such as photons, electrons, and ions make up the bulk of dark matter?

Electrons and ions are charged particles. Charged particles radiate light, and, therefore, are the antithesis of dark matter.

Photons, on the other hand, are massless. Nonetheless, although energetic photons do have a mass equivalency, photons give expression to light, and, therefore, once again, are the antithesis of what one would expect dark matter to be.

Another possible candidate for dark matter involves a particle that exists in stupendous quantities. This is the neutrino.

On the surface of things, neutrinos are an attractive candidate for dark matter because the former particles carry no electrical charge, are nearly massless, and interact with matter only very weakly. Therefore, neutrinos appear to give expression to many of the same features that seem to characterize dark matter.

Billions of neutrinos pass through us every second of the day. Thus, despite their very tiny masses, if neutrinos were sufficiently numerous, then, perhaps, they might constitute the dark matter of the universe.

Evidence for the possible existence of something like neutrinos began with the study of beta radiation, one of three kinds of radiation associated with radioactive decay (the other two being alpha particles and gamma rays). Beta radiation consists mostly of energetic electrons, but there were hints that something else might be present as well.

Scientists expected that in any instance in which a given radioactive element decayed into a lighter element, then, the beta radiation that is generated during the process of decay should be the same since according to Einstein's famous equation, $E=mc^2$, the amount of energy released in the form of that radiation should be proportional to the amount of energy lost during the decay process. However, scientists discovered that the amount of energy present in beta radiation varied.

In addition, researchers found that the amount of mass involved in the decay process did not seem to be conserved. Something appeared to be missing.

Wolfgang Pauli proposed in 1930 that, maybe, some portion of the missing energy in beta decay might have been carried away by a heretofore unknown particle. Moreover, if the missing energy were – for unknown reasons -- apportioned differently with each release of beta radiation, then, this might be able to account for the energy variability that had been observed taking place in conjunction with beta radiation.

Four years later, in 1934, Enrico Fermi developed a more complete theoretical account of beta radiation. Fermi theorized that when some given heavier atom decayed into a lighter atom, this was due to the transition of a neutron into a proton … a transition that was accompanied by the release of an energetic electron and one of Pauli's ghost particles that Fermi called "neutrinos" (meaning: "little neutral ones") – i.e., beta radiation.

Fermi conjectured that the neutron to proton transition suggested that some sort of underlying force was present. He did not believe this force was electromagnetic in character, because if this were the case, then, beta radiation should have been present in amounts billions of times more than were observed, and, therefore, he believed the cause of the neutron to proton transition during radioactive decay that was accompanied by beta radiation was due to some much weaker and unknown force that, at the time, was referred to as Fermi interaction.

| Cosmological Frontiers |

The existence of Fermi's interaction was proven in 1983 with the discovery of the W and Z bosons that carry the weak force. However, scientists also discovered that there was more than one type of neutrino.

Pauli had proposed the existence of a particle like the neutrino in 1930, but Clyde Cowan and Frederick Reines did have their first encounter with a neutrino – the electron neutrino -- until 1956. The muon neutrino was theoretically predicted during the latter part of the 1940s and was empirically uncovered by Leon Lederman, Melvin Schwartz, and Jack Steinberger in 1962, while the Tau neutrino was theoretically predicted in the 1970s, and was empirically detected through the DONUT (Direct Observation of the NU Tau) project at Fermilab in 2000.

As briefly discussed in 'Chapter 2: Antimatter Asymmetry' in *Volume II* of *Final Jeopardy: Physics and the Reality Problem*, neutrinos apparently oscillate ... that is, change into one another (at least, this is one theory). According to quantum mechanics, the rate at which particles oscillate is a function of their masses.

The more that the masses of two oscillating particles are similar to one another, the more slowing such oscillation will take place. On the other hand, within limits, the more disparate the masses of two oscillating particles are, then, the more quickly such oscillations are likely to occur.

The three neutrinos seem to have masses that are sufficiently different from one another to make a relatively rapid process of oscillation possible and, therefore, detectable. Because neutrinos: Had mass; were electrically neutral; interacted only very weakly with the sort of matter with which we are familiar; were stable (that is, did not decay into still lighter elements), and were quantitatively numerous, neutrinos were considered by many researchers to constitute a good candidate for being dark matter.

In dark matter research, particles with the properties of neutrinos are known as a WIMP (Weakly Interacting Massive Particles). If other kinds of WIMP particles beside the neutrino exist, they have not yet been detected.

Since dark matter has been calculated to comprise some 27% of the 'stuff' of the universe, and since, as far as has been determined, dark matter only gravitationally interacts with visible matter (which comprises 4% of the 'stuff' of the universe), dark matter is likely to have a significant impact on how the universe is, and becomes, structured. Computer simulations have been conducted using theoretical WIMP candidates exhibiting different velocities in order to determine whether any of those simulations were capable of reflecting the sort of structure we see in the universe today.

Slow-moving WIMPs yielded better results in such simulations (that is, generate results that more closely resemble what is observed in the universe around us) than fast-moving WIMPs did. Slow-moving particles permit (via gravitational accretion) dense structures like the ones that are observed in the universe today to form more quickly than fast-moving particles did.

Slow-moving dark matter is referred to as "cold dark matter". Neutrinos travel at near-light speeds and, therefore, if they do constitute a form of dark matter, it would be hot dark matter.

In simulations, hot dark matter will, eventually, lead to the formation of large-scale, dense structures. However, the time required for such structures to arise is out of whack with the estimated age of the universe, and, therefore, a universe shaped by hot dark matter would take too long to develop the structures that we see currently see in the universe.

As far as the Standard Model of quantum physics is concerned, the only remotely viable WIMP candidate in sight is the neutrino. Since, at best, neutrinos are a form of hot dark matter, and given that hot dark matter does not seem to be able to generate the structures we see today within the timeframe given by the presently estimated age of the universe, then, the Standard Model has no ready answers for identifying the nature of dark matter.

What about beyond the horizons of the Standard Model? Are there any dark matter candidates to be found there?

One possibility involves the issue of symmetry. In physics, symmetry exists when some property, principle, or law remains invariant despite the presence of transformations that take place involving such properties, principles, or laws.

Einstein's special and general theories of relativity constitute symmetries. Each framework, in its own way, permits physical transformations (involving moving bodies and gravitational events respectively) to occur while simultaneously preserving the laws of physics.

Gauge symmetries are also very important in physics. For instance, gauge symmetries exist when one can show that irrespective of how scales might be set in a given system, the behaviors that occur within such a system operate in accordance with the same set of laws.

Moreover, particles in systems that operate in accordance with gauge symmetry oftentimes are constrained in certain ways. For example, if a photon were not massless, then, the gauge symmetry to which electromagnetism gives expression would be broken, and, consequently, the gauge symmetry of electromagnetism imposes certain constraints on the properties of the underlying boson or carrier of force -- namely, the photon.

During the 1970s, a group of scientists in the Soviet Union began to explore a new kind of symmetry. This symmetry involved the relationship between fermions (particles – such as quarks and leptons with half-integral spin that obey the Pauli exclusion principle) and bosons (particles -- such as the photon, gluon, W and Z particles of the weak force, the Higgs, and the hypothetical graviton -- that have integral spin and do not obey the Pauli exclusion principle)

The symmetry being explored tied forces and matter together, making them interdependent. More specifically, the theory proposed that for each fermion there was a corresponding boson and vice versa.

The counterpart particles were referred to as superpartners. The theory came to be known as Supersymmetry.

The superpartner of a given fermion or boson should give expression to the same sort of mass, amount of electrical charge, and other properties that are possessed by their Standard Model counterparts. At least, this should be the case if the symmetry between particles and their superpartners were perfect in nature.

Apparently, however, if Supersymmetry turns out to be a viable account of the quantum world, that symmetry will not be perfect. It will contain broken symmetries of one kind or another.

If broken symmetries were not present in Supersymmetry – assuming, for the moment, that the underlying theory is viable -- researchers in high-energy physics already should have come across evidence of a superpartner (i.e., the selectron) that has the same mass as an electron while displaying other properties that distinguish selectrons from electrons. Yet, to date, no one has detected the presence of a selectron – or any other superpartner – in the particle debris that is generated through high-energy accelerators and colliders.

The proponents of Supersymmetry interpret the foregoing absence of evidence as merely being an indication that superpartners must be heavier than initially thought. How much heavier is unknown, and, therefore, detecting their presence becomes a difficult task (assuming such heavier entities exist).

Discovering the Higgs boson presented a similar challenge because, among other things, no one was quite sure what its mass was. Past and present failures to detect the presence of superpartners could be due to the unknown mass issue, or such failures might just be an indication that those superpartners don't exist.

Despite the absence of evidence – at least thus far – to support the idea of Supersymmetry, physicists are reluctant to give up on that theory. This is because if Supersymmetry – or something that plays a similar role – is not correct, then, the Standard Model is beset with some critical problems.

For example, according to quantum physics, the relatively recently discovered Higgs boson is a telltale sign of the presence of a Higgs field through which particles that are sensitive to its presence acquire mass. Particles like the photon do not possess the foregoing sort of sensitivity, and, as a result, are massless.

As critically explored previously (both in this volume as well as in *Volume II* of *Final Jeopardy*), quantum physics maintains that virtual particles are continuously popping into and out of existence. If this is the case, then, the Higgs boson should interact with those virtual particles and, in the process, become very heavy ... perhaps trillions of times its actual mass.

The Higgs boson has been found to have a mass of 125.09 GeV. So, if virtual particles operate in the way that quantum theory maintains, what constrains the interaction between the Higgs boson and virtual particles and, in the process, keeps the Higgs boson from becoming incredibly heavy?

Proponents of Supersymmetry argue that Standard Model virtual particles and their virtual superpartners will cancel each other out. This would bring about stable masses for particles like the Higgs boson.

If Supersymmetry is not a correct account of quantum mechanics (and, so far, there is no evidence that it might be) then, what constrains how non-virtual particles interact with virtual particles in order to be able to generate the stable masses that are observed? Of course, one way of engaging the foregoing question is to entertain the following possibility: To whatever extent virtual particles actually exist, they do not exist in quantities that are anywhere near what quantum theory supposes is the case.

Many physicists appear to believe that the existence of virtual particles necessarily follows from Heisenberg's uncertainty principle, That is, because one cannot know the precise amount of energy that is present in a system, this leaves room for the possibility that particles could be generated spontaneously as long as they become extinguished soon thereafter ... time and energy are said to be conjugate variables.

Whatever the uncertainties of measurement might be, this is a reflection of the process of measurement and doesn't necessarily have anything to do with the nature of the facet of ontology that is being measured. The fact there are uncertainties – due to problems inherent in the measurement process – does not mean that ontology is capable of spontaneously generating and extinguishing virtual particles beneath the curtain of methodological uncertainty.

The precise amount of energy that is present in a system -- but which cannot be pinned down due to methodological limits -- might permit modes of "virtual" manifestation to take place within certain parameters. However, while quantum theory might contend that such methodological unknowns are capable of housing an infinite set of possibilities involving virtual activity, the actual ontology of such a system might be quite different.

Theoretically, virtual activity is calculated to give expression to a vast array of possibilities. Ontologically, the Higgs boson has a stable mass, and, consequently, there is nothing to rule out the possibility that the reason why reality – as opposed to theory – is able to manifest itself in such a constrained way is because the virtual particle activity predicted by theory is not taking place ... either at all or to an extent that is anywhere near the levels that are presupposed by quantum theory.

Virtual particles might be something of a red herring-like issue as far as Supersymmetry is concerned. If virtual particles – to whatever extent they actually exist – do not operate in the way, or to the extent, that theory requires, then, there is no need for Supersymmetry to rush to the rescue because the stability of particles like the Higgs boson can be explained in other, much simpler ways, as indicated during the discussion of the last three paragraphs .

Some versions of Supersymmetry predicated that protons decay. No evidence has surfaced yet indicating that protons do, in fact decay – at least not at the rate hypothesized, and, therefore, for the present time, such models of Supersymmetry have been put aside.

There is another edition of Supersymmetry that considers protons to be stable, and, therefore, they do not decay. Such models are said to possess R-parity symmetry in which, among other things, superpartners are generated or extinguished in pairs

However, in order for R-parity to be observed, the lightest superpartners must exhibit stability. They cannot decay into still lighter particles.

The simplest Supersymmetric models entailing R-parity symmetry contain seven possibilities that might satisfy the conditions necessary to qualify as dark matter candidates. Three of these candidates are the superpartners of neutrinos – namely, sneutrinos -- while the remaining four possibilities are superpartners for: Two kinds of Higgs bosons, the Z boson, and the photon (known, respectively, as Higgsino, zino, and photino).

The latter four possibilities are collectively referred to as neutralinos. In order for Supersymmetry models that observe R-parity symmetry to be viable, the lightest of the superpartner candidates must be stable.

The problem is that researchers have had difficulty demonstrating which of the neutralinos might be lightest. A fair amount of this difficulty is due to the fact that the lightest superpartner candidate could be any one of a number of combinations involving the aforementioned four superpartners.

Different combinations of neutralinos give expression to different interactional properties. Just as the interactional properties of a photino are not the same as those of a zino, so too, the interactional properties of one combination of neutralinos will not be the same as other possible combinations.

A great deal of research has been conducted in conjunction with neutralinos. So far, however, trying to figure out: What the lightest neutralino is, whether it is stable, and whether such a possibility -- if it exists – fits the bill for dark matter, has proven to be a difficult row to hoe.

Various possible dark matter candidates have been eliminated from consideration. However, the search for viable candidates continues.

One such project is known as: Super Cryogenic Dark Matter Search. The first stage of the experimental project is taking place in the Soudan Underground Laboratory in Minnesota (about a half mile below the surface), and a later stage of that same project is scheduled to take place in the deeper SNOLAB facility in Sudbury, Ontario.

One of the central problems confronting such searches involves finding ways to eliminate possible sources of noise or signal contamination so that one can detect the presence – if it exists – of dark matter. The aforementioned project creates an environment that is just above absolute zero (perhaps $1/100^{th}$ of a degree Kelvin above absolute zero or closer), and, then, uses additional methods of shielding to block out still other possible sources of noise that might be able to contaminate or dilute the signal being sought.

The CDMS project also uses detectors known as iZIP (interleaved Z-sensitive Ionization Phonon) that are made from extremely thin superconducting films that have been deposited on germanium crystals. Such detectors are intended to measure the recoil energy that results from the kinds of ionization that researchers believe will characterize the interaction of a WIMP with the nuclei of "normal" matter (Obviously, if dark matter exists, then percentage-wise, dark matter would constitute the norm, and Standard Model matter would be abnormal).

The SNOLAB in Sunbury will be even better protected from possible sources of contaminating noise than the facility in Minnesota is. For example, because of its deeper location beneath the surface of the Earth, the Sudbury stage of research will be better protected against the impact of cosmic rays.

Interpreting the data that arises in conjunction with such instruments is fraught with problems. Sometimes researchers have difficulty understanding the significance, if any, of the data that is recorded.

For example, as the first part of the aforementioned CDMS project was coming to an end and thoughts began to turn toward the process of replacing CDMS by its successor – SuperCDMS – two events took place. On August 5, 2007 and again on October 27, 2007 "something" had tripped the detectors.

Five events of a specified kind were considered necessary to make a plausible case that something of statistical significance had occurred. Since only two events had taken place, the meaning of those events was unclear.

Some non-dark matter form of radiation or cosmic ray could have triggered the detectors. On the other hand, the two events might have been harbinger signals that marked the presence of dark matter.

Project participants began examining the events in detail. They wanted to determine what the quality of their data might be.

After performing more than fifty checks concerning those events, the CDMS researchers came to the conclusion that the quality of the data was very high. In other words, they had not been able to come up with anything during their checking process that would have allowed them to dismiss such events as being more likely to have been the result of background noise, instrument malfunction, coincidence, and the like than, possibly, having been the result of an encounter with dark matter.

The quality of the foregoing events served to enhance the possibility that the events might have involved dark matter. However, the statistical significance of that data forced the researchers to place a number of cautionary flags around the quality such events.

The CDMS researchers were faced with the prospect of either accepting a possibility that might be false or rejecting data that actually might have reflected the presence of dark matter. There was no way to tell which possibility was correct.

There are other research programs besides CDMS/SuperCDMS that are pursuing the task of detecting dark matter. Those projects are taking place in different parts of the world and are using different mediums (e.g., xenon, argon, and neon), cooled to different temperatures, in an effort to detect the presence of dark matter.

The foregoing experimental programs are fairly expensive to set up and operate. However, there are projects that have been devised that cost only a small fraction of the foregoing kinds of programs.

For example, consider one research project involving axions. An axion is a hypothetical form of neutralino that is a trillion times lighter than an electron.

Unlike various WIMP candidates, axions are not massive, and, moreover, they are considered to be unlikely to interact with baryonic matter. However, axions do possess a property that is sensitive to the presence of magnetic fields.

If a magnetic field is sufficiently strong, then axions that encounter it tend to disintegrate. This disintegration produces a photon.

If a detector could be built that prevented such photons from escaping, then those products of disintegration would bounce about and generate a microwave signal. The microwave signal would be a detectable sign – theoretically -- that an axion disintegrated and produced a photon.

In 1983, Pierre Sikivie envisioned a radio-receiver-like design to detect the foregoing kinds of signals. Fourteen years later, in 1997, two other researchers – Karl van Bibber and Leslie Rosenberg – put together a axion detector prototype.

Axion-related research of the foregoing kind is referred to by the acronym ADMX. This stands for Axion Dark Matter eXperiments.

So far, none of the aforementioned research projects – whether expensive or cheap -- has been able to uncover the, so far, hidden identity of dark matter. As the technology surrounding the issue of detection becomes more sophisticated, one, or another, of the foregoing projects might be able to announce that the elusive quarry finally has been corralled.

On the other hand, if research projects like ADMX or CDMS/SuperCDMS continue to produce null results, then at some point researchers face the following questions. Do null results indicate: (1) The research project just hasn't been able to formulate the right search parameters and/or doesn't have sufficiently sensitive (or the right kind of) detection instruments to be able to detect the presence of dark matter; or, (2) the signal they are seeking does not actually signify the presence of dark matter and, therefore, even if found, will not bring one any closer to understanding the nature of dark matter; or, (3) what the researchers are looking for signs of – namely dark matter -- doesn't exist?

Some physicists believe the presence of dark matter might be detectable through the products that are created during processes of annihilation.

If R-parity holds in a given model of Supersymmetry, then one of the possibilities is that if two neutralinos were to collide and there were an even number of superpartners present, the resulting annihilation would leave behind Standard Model matter plus energy, but such a debris field could be used as a means of indirectly detecting the presence of dark matter.

For example, Supersymmetry models possessing R-parity predict that the photons arising from the annihilation of dark matter will generate a certain kind of gamma ray. When such gamma rays enter the Earth's atmosphere, a debris field is created, and the latter field could be used to infer the existence of the sort of gamma rays that are theoretically predicted to arise during the process of dark matter annihilation.

In addition, when dark matter undergoes annihilation, then, in addition to signature sorts of photons being produced, certain kinds of dark matter-related antiparticles and neutrinos are also predicted to arise. In each case, signals indicating the presence of dark matter have to be differentiated from background noise, and doing so involves a variety of difficulties.

Research has been conducted in all of the foregoing areas. However, to date, there is no data that has been collected pointing unmistakably – and in ways that can be replicated -- to the presence or existence of dark matter.

Science, however, does not progress just by discovery. It also proceeds through eliminating possibilities.

For example, as noted previously, there are seven possible dark mater candidates that are connected with Supersymmetry models involving R-parity. Three of those candidates involve the superpartners of neutrinos known as sneutrinos.

If sneutrinos existed, many physicists believe that projects like CDMS should have been able to detect vast quantities of them by now. Since this has not occurred, many physicists have been inclined to cross sneutrinos from the list of viable dark matter candidates.

Moreover, projects like CDMS are predicated on the assumption there are methodological means that can be developed through which WIMP candidates will reveal their presence (and nature) by interacting with Standard Model matter in ways that can be detected. Thus, CDMS is attempting to measure the amount of energy recoil that has been hypothesized to occur when a WIMP entity interacts with the nuclei of so-called "normal" matter.

However, what if -- with one exception -- weakly interacting massive particles do not interact with Standard Model at all? What if dark matter interacts with Standard Model matter only through the force of gravitation?

One might be able to determine that dark matter and Standard Model matter give rise to either the same kind of, or different sorts of, gravitational behavior, but, perhaps, not much more than that. Under such circumstances, even if one were able to distinguish between dark matter and Standard Model matter through their respective gravitational signatures (assuming there is one), determining the complete nature of dark matter might continue to prove a very hard nut to crack.

Experimental issues aside, there are many exotic theories and models that seek to provide insight into the nature dark matter. Some of these theories and models are rooted in the notion of extra dimensions.

The original version of string theory emerged in the 1960s as an attempt to bring some sort of order to the particle zoo that existed at that time. Strings gave expression to different modalities of vibration, and, conceivably, the array of particles that had been discovered in the 1960s could be unified through the idea that the apparent differences among particles was a function of the way in which strings vibrated under various conditions.

Although the idea of strings was soon overtaken by the rise and development of quark models, there was one aspect of string theory that was appealing ... at least to some researchers. More specifically, string theories predicted the existence of additional particles, and while some investigators felt that such particles served to undermine the viability of string theory, other individuals noted that mixed in with those additional particles was one that had properties very much like the hypothetical graviton particle in models involving quantum gravity.

The tide of interest in string theory remained at low ebb for several decades. However, that tide began to rise in 1984.

In that year, Michael Green and John Schwarz introduced a theory of strings that was super in several respects. Not only did their theory overcome many of the problems that previously seemed to plague the initial versions of string theory (for example, the issue of tachyons – particles that, supposedly, were able to move faster than the speed of light and back through time), but as well, their newly formulated theory contained elements that could be used to describe the four known forces of physics, and, therefore, alluded to the possibility that string theory might lead to the long sought holy grail of physics ... a unified theory of physical reality.

However, in order for the new Supersymmetric string model to work, it had to possess a number of dimensions beyond the usual four with which we are familiar. The revamped version of string theory led to infinities and inconsistencies if it did not operate in either 10 or 26 dimensions, and, more importantly, only ten-dimensional versions of string theory actually provided viable ways to describe the properties of fermions and bosons that had been established through the Standard Model of quantum physics.

Given that we are only familiar with four dimensions, where were the other six dimensions? In 1926, Oskar Klein proposed an idea that explained how dimensions could exist that were not visible and, therefore, might be applicable to the new theory of strings.

Klein referred to such existent, but non-visible, dimensions as being wrapped or folded up. The modern term is "compactification".

These folded up dimensions were exceedingly miniscule. In fact, they were so small that the realm of quantum physics would play a major role in determining how particles traveled through such dimensions and what effect journeying through that kind of rolled up dimension would have on the properties of those particles.

Energy is one of the properties that – at least theoretically – is affected by journeying through compactified dimensions or spaces. The smaller a rolled up dimension is, the smaller the wavelength associated with a particle traveling through such a space will have to be, and the smaller that wavelength is, the more energetic such a particle will be.

Furthermore, if the wavelength of a particle is able to fit into such a compactified space and if, in addition, that wavelength were sufficiently small, then, a particle would have certain degrees of freedom to move about in that rolled up dimension. Such movement would give expression to kinetic energy.

To an outside observer, particles moving about, and traveling through, a compactified space or dimension would appear to be very energetically heavy or massive. Perhaps, such particles might be good candidates for dark matter.

The foregoing possibility is known as a Kaluza-Klein state. In order for such states to be able to continue to exist, they would have to satisfy the law of conservation of momentum -- that is, they would be unable to transfer their momentum to anything else within that compactified space.

If such Kaluza-Klein states do exist, and if those states do give expression to dark matter, then, dark matter is not about mass per se. Rather, dark matter becomes a function of what occurs when a particle of some kind is trapped in compactified space in the form of a Kaluza-Klein state, and the particle's wavelength, together with its momentum, give rise to the mass-like properties of dark matter (i.e., it would be sensitive to gravity's presence).

If one is able to adjust parameters -- such as the size of the compactified space through which a given particle travels, as well as the wavelength and movement of such a particle within that sort of rolled up space/dimension -- one can generate an array of theoretical possibilities that might help to provide some computational traction in relation to the idea of dark matter. Consequently, in many ways, being able to adjust parameters to generate the kinds of values one needs to make something like a Kaluza-Klein state version of dark matter work is a perspective that resonates, to some degree with the epicycles that were continuously introduced into the Ptolemaic system in order to make various predictions concerning the movement of heavenly bodies work out.

The foregoing sorts of additions might have led to computations and predictions that solved certain problems. However, that success often came at a price – namely, the obfuscation that arose in conjunction with trying to understand what actually was taking place.

There also seems to be a bit of equivocating going on when some individuals treat extra dimensions as variables that can be used to describe certain kinds of behavior, while other individuals treat dimensions as if the latter notions were real, physical spaces through which particles can journey. If dimensions are merely variables to be taken into consideration when trying to account for behavior, then, one doesn't necessarily have to explain them away as being compactified and, therefore, not physically visible to us.

Notwithstanding the foregoing considerations and assuming that dark matter actually exists in the form of some kind of WIMP (weakly interacting massive particle) -- and assuming that the phenomenon known as "dark matter" is not due to some other process such as, say, the electromagnetic effects present in plasma dynamics that, apparently, taking place within and between galaxies – then, there are some additional cosmological problems that are entailed by the idea of dark matter. For instance, if dark matter actually exists, what role did it play, if any, in the Big Bang (assuming, of course, that the Big Bang did take place)?

If dark matter interacts with Standard Model matter through non-gravitational means, then, presumably, such interaction might affect how the primal singularity formed (if it did) or how Big Bang took place (if it took place). If dark matter and Standard Model matter do interact, then how weak is that interaction and what is the nature of that interaction and what impact, if any, would such interaction have on the properties of the Big Bang?

Assuming that the Big Bang took place, would dark matter particles have generated heat in the same way as Standard Model particles supposedly did with respect to helping to generate the very high temperatures that usually are associated with the primal singularity? Is it possible that the presence of dark matter might have affected the primal singularity's temperature (upwards or downwards) in some fashion?

Is it possible that dark matter particles do not operate in accordance with quantum mechanics ... partially or wholly? If dark matter does not operate in accordance with the principles of quantum dynamics, then, how would this have affected the properties of the alleged primal singularity, or the Big Bang, or what ensued following the Big Bang?

Does dark matter contribute in any way to Cosmic Microwave Background Radiation? If it does, how is this reflected in the data generated through COBE (Cosmic Background Explorer) or WMAP (Wilkinson Microwave Anisotropy Probe)?

If dark matter doesn't contribute to CMB, then do either COBE or WMAP provide an accurate picture of how the universe subsequently unfolded? The foregoing question is particularly pertinent since dark matter supposedly is five times more plentiful than Standard Model matter is and, as a result, dark matter, presumably, had a disproportionate impact on how large-scale structures developed across the universe relative to the contributions that Standard Model matter might have (unless, of course, the electromagnetic dynamics of plasmas generated by Standard Model matter shaped the structural development of the universe more than most astrophysicists currently believe was the case).

The facets of anisotropy that have been detected in the COBE and WMAP data have been attributed to quantum fluctuations in temperature as the universe cooled down following the Big Bang. Is it possible that such anisotropy might – partially or wholly -- be an indirect reflection of the presence of dark matter rather than quantum fluctuations?

How does one differentiate between whether a given piece of COBE or WMAP data is due to quantum fluctuations and/or the presence of dark matter? If we don't know what the nature of dark matter is, then, how does one disentangle the contribution of dark matter and quantum fluctuations when interpreting the COBE and WMAP data?

If the composition of the primal singularity (the alleged progenitor of the Big Bang) consisted of more than five times as much dark matter as Standard Model matter (and assuming that such a singularity existed), was the dark matter distributed uniformly or irregularly throughout that singularity? If the foregoing distribution were irregular, how would that have affected the properties of the Big Bang ... if at all?

One can, of course, take a look at the universe today and claim – rightly or wrongly – that matter is distributed throughout the universe in a roughly homogenous manner. Therefore, on that basis, on might conclude that the distribution of matter in the primal singularity must have been uniform and/or that the nature of the Big Bang brought about such a uniform distribution.

However, there is considerable evidence indicating that matter – both dark and visible – is not distributed homogenously throughout the universe. Consequently, determining the extent to which the present state of the universe is capable of shedding light on the distribution of both Standard Model matter as well as dark matter in the primal singularity is not necessarily a straightforward issue.

Many scientists maintain that if viewed from an appropriate scale, then the universe exhibits isotropy and homogeneity. Nonetheless, isn't it possible that such a perspective is just an artifact of a methodology of scaling that pushes the so-called "observational" process to a point where galaxies, galaxy clusters, cosmic walls, and voids, are reduced to mere data points that can, in the abstract, appear to be relatively uniform and homogeneous if one increases one's viewing scale by a sufficient number of orders of magnitude?

However, if the universe is not infinite, then, isn't there a limit to the kinds of scales through which one can view the universe? If there are such limits, then isn't it possible that when considered in its own terms (rather than through the filters of scaling methodology), the universe is neither isotropic nor homogeneous?

Does space interact differently with dark matter than it does with Standard Model matter? Does space interact at all with either dark matter or Standard Model matter?

If dark matter is not sensitive to the presence of electromagnetism, the weak force, or the strong force, then is it possible that dark matter might, nonetheless, interfere with, block, or attenuate the dynamics of those forces in some way? In other words, is it possible dark matter gives expression to a force field of sorts (or something like it) that cannot be penetrated – or penetrated only very weakly – by the forces (other than gravity) that govern Standard Model matter?

If dark matter and Standard Model matter both comply with the equivalence principle of general relativity, why do galactic halos appear to consist largely, if not exclusively, of dark matter? Does dark matter interact with gravitation in the same way that Standard Model matter does? If not, what are the differences?

Why does dark matter appear to give preferential gravitational consideration to dark matter? Why does Standard Model matter seem to give preferential gravitational consideration to Standard Model matter?

| Cosmological Frontiers |

Could black holes be comprised of various amounts of dark matter, and, if so, how would the presence of that dark matter affect the properties of those black holes? Is it possible that some black holes might consist entirely of dark matter, and, if so, would such black holes exhibit different kinds of behavior than a black hole consisting of Standard Model matter or a mixture of the two kinds of matter?

The issue of dark matter raises many fundamental questions involving: The nature of the primal singularity, as well as the dynamics of the Big Bang, in particular, and the properties of cosmology in general. Therefore, even if one were to assume that the Big Bang did take place, one would not necessarily be able to move very far in any conceptual direction – at least not with any degree of confidence that was warranted – that would be capable of providing a plausible, verifiable explanation concerning how the primal singularity formed, or how the Big Bang – or ensuing cosmological events -- unfolded over time.

If we don't understand the initial conditions of the primal singularity – including the presence, status, properties, and role of dark matter – and if we don't understand the dynamics of the Big Bang and what role, if any, dark matter played in that Big Bang, then whatever conclusions one draws concerning how the universe got to be the way it is today is purely speculative. Such conclusions are little more than hermeneutical narratives that weave together a few empirical strands that are known with many more strands that are unknown and, then, seek to pass themselves off as scientific tapestries despite the absence of any real concrete proofs amidst the many colorful possibilities to which such narratives give expression.

For the most part, the existence of dark matter is predicated on the difference between the amounts of mass that have been calculated to be necessary to explain the rotational motion of galaxies and the amounts of visible matter that have been estimated to exist in such galaxies. The matter that remains after all of the visible matter has been taken into account is referred to as dark matter.

There are alternative ways of trying to account for the rotational motions of galaxies that do not involve positing the idea of dark matter. For example, leaving aside the issues that were discussed in Chapter 4 ('Electric Universe') of this book, Mordechai Milgrom, an

Israeli physicist, developed a variation on Newtonian physics that is referred to as MOND (MOdified Newtonian Dynamics).

MOND proposes there are conditions under which Newtonian physics breaks down. These conditions have to do with the acceleration exhibited by an object such a star or a galaxy.

According to Milgrom, if the acceleration of star or galaxy is greater than $1.2 \times 10^{-8}$ centimeters per second squared, then Newtonian dynamics works perfectly well, and, consequently, one does not have to invoke dark matter to account for rotational motions. However, if the rate of acceleration of a star or galaxy falls below the foregoing threshold, then, the force of gravity appears to change.

Below the aforementioned threshold acceleration value, gravity no longer falls off inversely with the square of the distance. Instead, gravity appears to decrease as a function of just distance.

In addition, when the acceleration of a star or galaxy falls below the critical threshold, the force of gravity is no longer proportional to the product of the masses times the gravitational constant. Rather, below the critical threshold, MOND claims that the Newtonian force is proportional to the square root of the mass times the gravitational constant.

Portions of a galaxy that occupy orbits that accelerate at rates above the aforementioned critical level will behave in ways that can be accurately described through Newtonian dynamics. But, if there are portions of a galaxy occupying orbits whose rate of acceleration falls below the critical rate of acceleration, then, according to Milgrom, Newton's dynamics have to be modified in the ways indicated by MOND.

MOND focuses on Newtonian dynamics. It does not cover the sort of specialized dynamics that are handled by Einstein's theory of general relativity.

Although MOND is not universal in its sphere of applications, nonetheless, it still is capable, under some circumstances, of making reliable calculations concerning the rotations of galaxies. The descriptions of Newton dynamics and general relativity tend to agree when making calculations involving rotational motions of galaxies, and, therefore, within this context, MOND is departing from the sort of predictions and calculations that would be made through the frameworks of both Newton and Einstein.

Within certain limits, MOND is able to do a fairly good job of describing the rotational motion of galaxies (especially those that are known as low-surface-brightness galaxies). However, in other cases (such as when dealing with large clusters of galaxies), although MOND is able to account for a portion of the seemingly anomalous rotational motions of some galaxies, nonetheless, there appears to be considerable mass that is unaccounted for and that needs to be taken into consideration when attempting to generate plausible accounts for the rotational motions of galaxies that belong to large clusters of galaxies.

Furthermore, while MOND enjoys a certain degree of success when it is used to describe the motions of stars within galaxies – in fact, it tends to perform better than dark matter-based models do with respect to such descriptions -- MOND is far less successful when used to describe the dynamics that take place outside of galaxies. Theories that incorporate the idea of dark matter fare much better than MOND does when it comes to providing descriptions and predictions that usefully reflect observed extra-galactic dynamics.

In addition, MOND does not provide an explanation for why a change in the rate of acceleration should bring about a shift in the behavior of gravity. That is, why should a force that, generally speaking, is inversely proportional to the square of the distance, as well as is proportional to the product of masses times the gravitational constant, suddenly change and become inversely proportional to just the distance as well as become proportional to the square root of the masses times the gravitational constant?

Of course, it is possible that the presence of dark matter might slow down rates of acceleration. If this were the case, then, the acceleration threshold merely indicates the point at which dark matter enters the picture and begins to shape gravitational behavior.

Conceivably, under the foregoing set of circumstances, dark matter might exhibit a different kind of gravitational force than Standard Model matter does. If this were the case, then the principle of equivalence might not hold <u>collectively</u> for both dark matter and

Standard Model matter but still could hold for each kind of matter considered separately or individually.

In other words, dark matter might attract other dark matter in compliance with the principle of equivalence, just as Standard Model matter attracts other Standard Model Matter in compliance with the principle of equivalence. Nonetheless, dark matter and Standard Model matter might not attract one another in compliance with the principle of equivalence, and this might account for why dark matter and Standard Model matter often appear to clump together somewhat independently of one another.

If the foregoing were correct, then, maybe, MOND is, to some degree, a reflection of some of the differences in gravitational behavior that exist with respect to dark matter and Standard Model matter. When dark matter is present, perhaps gravity operates somewhat differently than when Standard Model matter is present.

Theorists have proposed that there might be more than one kind of Higgs field and accompanying Higgs boson. Similarly, perhaps, there is more than one kind of gravitational field, each with a different kind of graviton governing the properties of gravitational attraction involving dark matter and Standard Model matter.

There is one other aspect of MOND that is either very interesting (in a significant way) or, perhaps, just coincidental. The critical threshold of acceleration that is at the heart of MOND – namely, $2.1 \times 10^{-8}$ cm/sec$^2$ – resonates with the value that has been calculated for the rate at which the universe is supposedly expanding – namely, $10^{-8}$ centimeters per second.

Does the foregoing resonance indicate there might be some sort of connection between dark matter and dark energy (especially if MOND turns out to be an indication – contrary to the intentions of its inventor -- that there are gravitational differences between dark matter and Standard Model matter)? Perhaps, but like so many things in cosmology, at the present time, there are just too many unknown dimensions of the universe to arrive at any firm conclusions concerning such an issue.

There is, at least, one set of empirical data that might, or might not, carry implications for the possible links between dark matter and dark energy. The empirical data were generated through Pioneer 10 and Pioneer 11 spacecrafts.

The primary purpose of Pioneer 10 and 11 was directed toward probing the outer planets of the solar system. Pioneer 10 was launched in 1972 and Pioneer 11 followed suit the next year.

Once the two space vehicles had completed their primary mission, they were to continue moving away from the sun in opposite directions along the plane of the solar system. The Jet Propulsion Laboratories in Pasadena, California continued to monitor the velocities and trajectories of the two vehicles via Doppler shifts.

Predictions were made concerning certain aspects of the future trajectories of the two ships. These predictions tried to take into account the gravitational impact that various occupants in the solar system might have on those trajectories.

The predicted trajectories did not correspond with the actual trajectories of the two ships. The discrepancy was $8 \times 10^{-8}$ centimeters per $sec^2$ and seemed to be the result of some kind of accelerating force that was tugging on the two vehicles from the direction of the Sun.

Researchers who are independent of JPL have gone over the data. All manner of possibilities (gas leaks, etc.) have been taken into consideration in an attempt to account for the foregoing discrepancies, but, to date, no one has been able to identify any plausible source of error that could account for the differences between what had been predicted and what was observed in relation to the trajectories of the two space craft.

If the Pioneer-trajectory data stands, one of the possible inferences that might be made concerning the significance of such data has to do with the nature of gravity. Conceivably, the properties of Newtonian gravity might operate slightly differently just outside of the solar system than they do within the solar system.

The difference between predicted and actual trajectory acceleration rates for Pioneer 10 and 11 was, as noted earlier, $8 \times 10^{-8}$ centimeters per $sec^2$. While that figure is roughly six times larger than the rate of acceleration threshold noted by Milgrom in conjunction with MOND, and while the Pioneer discrepancy is 8 times larger than the acceleration rate associated with an allegedly expanding universe, the fact that all of these figures are operating on the same scale of magnitude has raised more than a few eyebrows ... could it be that gravitation, dark matter, and dark energy are all related in some, yet-to-be-determined, manner?

Chapter 8: Expanding Horizons

Many astrophysicists apparently consider Alan Guth's notion of a universe that began inflating shortly after the Big Bang occurred to be quite attractive. None of them seems to know what the specific character of the phase transition was that gave rise to inflation, or why and how it decayed as quickly as it did, or why inflation had the precise properties that it did, or whether space is even something that is capable of being inflated, but the idea of inflation helped many physicists whistle their way past the cemetery of ghoulish problems that tended to surround the idea of the Big Bang (e.g., the nature of the singularity and why it exploded).

The phase transition underlying Guth's inflationary universe created a temporary vacuum. This vacuum, in turn, led to the generation of a negative pressure that expanded space exponentially.

Some individuals refer to the foregoing negative pressure as a form of gravitational repulsion. However, one wonders why those individuals would describe things in that way when, if such a negative pressure occurred, it wasn't actually a form of gravitational repulsion since only space seemed to have been affected by its presence.

Once cosmic inflation came to an end, then, the primary form of expansion taking place in the universe was due, supposedly, to the way in which the force of the Big Bang moved things apart from one another. In that context, scientists began to talk about the "flatness problem" – that is, whether there was sufficient mass in the universe to counter the force of expansion that had been set in motion by the Big Bang.

The relationship between mass and expansion was termed "omega". If there were enough mass in the universe to halt expansion but prevent such mass from, eventually, pulling back together into a Big Crunch, then, omega had a value of "1".

If omega was larger than "1", the universe eventually would collapse through the force of gravitational attraction. If, on the other hand, omega were smaller than 1 – say, .10 or .70 – then, the universe would go on expanding.

The values of omega that were greater than, or less than, 1 indicated how quickly or slowly things would happen. For values greater than 1, the more those values departed from 1, the faster eventual collapse would start to take place and, as well, accelerate toward complete collapse once such a reversal of expansion had occurred, while for values less than 1, then the closer such values were to 1, the more slowly the universe would continue to expand, but the smaller such values were, then this indicated that the universe would be expanding with a proportionate speed.

If omega deviated from 1 in any significant way – either above 1 or below 1 – then, the Big Bang would have been in trouble before it even had a chance to do anything interesting. For example, if omega were significantly greater than 1, then, the universe would have collapsed fairly abruptly (the greater the value of omega, the more abrupt the collapse would have been), but if omega were significantly smaller than 1, then the universe would have expanded so quickly that particles would never have been able to interact to develop the material complexities that we see in the universe today (the smaller the value of omega, the faster the rate of expansion would have been).

The fact that, on the one hand, after some 14 billion years the universe hasn't collapsed while, simultaneously, on the other hand, the universe permitted the growth of material complexity (thereby implying that the rate of expansion following the Big Bang could not have been excessive) indicates that the value of omega must have been fairly close to 1 as the universe began to get under way. If the universe had an omega value that was close to 1 at the beginning of things, then, it must have an omega value that is close to 1 now, and, if so, then perhaps, the only remaining question becomes a matter of determining how close to 1 omega is and what this means for the future of the universe.

Just as few, if any, individuals asked what Guth's inflationary universe might be expanding into, so too, few, if any, individuals seemed to be asking what the contents of the Big Bang were expanding into. Did space go on forever, or did it have determinate limits, and if such boundaries existed what were their properties?

Prior to Hubble, Lemaître, Gamow and the other proponents of an expanding universe (which implied the idea of a Big Bang), Einstein assumed that universe existed in some kind of steady state modality. However, after he released his initial version of the theory of general relativity in 1915, Einstein began to reflect on whether the universe would be able to remain in a steady state if the gravitational attraction of objects for one another began to play out over a period of time.

In 1917, Einstein introduced the value of Λ, lambda, to prevent the force of gravitational attraction from being able to overrun the universe. Λ referred to the energy density of space that was necessary to counter the force of gravitational attraction and, thereby, maintain a steady-state universe.

The foregoing energy density is not static. It constitutes a form of pressure that resists gravitational attraction.

Einstein had no idea what made such pressure possible or how it actually worked. However, whatever Λ was, it served his purposes – namely, to save the appearances of a steady-state universe.

Even without Λ, the universe might still have given expression to a steady state structure. The universe merely would have been a different kind of steady state universe than the one assumed by Einstein since, eventually, without Λ ,the universe would have become reduced – via gravitational attraction -- to some sort of high density, gravitationally powerful mass that might not do much more than persist as a steady-state blob.

More than a decade later, evidence was uncovered (by Hubble and others) indicating that the universe appeared to be expanding. As a result, Einstein rejected Λ and considered it to be the "biggest blunder" of his life ... indicating – at least implicitly -- that Λ had been little more than an arbitrary fudge factor intended to make his theory of gravity exhibit compliance with a particular theory of the universe (i.e., the steady state theory) that prevailed at the time Einstein introduced Λ into general relativity.

Sometimes the relationship between mathematical equations and reality is strange ... if not strained. General relativity purported to describe the behavior of gravitational reality, but Einstein's theory contained a term – namely, Λ – that, following Hubble's work, didn't appear to refer to anything and, yet, practitioners were reluctant to remove such a seemingly arbitrary element from the equations.

In any event, despite Einstein's acknowledgement that Λ constituted an embarrassing blunder, theorists and researchers who used the equations of general relativity set Λ to zero rather than removing Λ from general relativity altogether. Given Einstein's aforementioned opinion of Λ, holding on to Λ seems a rather bizarre thing to do, and one wonders why scientists didn't just banish Λ from consideration entirely... unless, out of respect for Einstein, they wanted to keep open some 'fudge factor' wiggle room to address – possibly -- certain kinds of unanticipated data that might show up in the future.

Returning to the issue of omega, an omega of 1 indicates that the universe is flat. An omega of 1 suggests that the universe has built into it a potential for equilibrium in which the force of expansion and the force of gravitation will be able to offset one another everywhere in the universe.

A lot of individuals felt uneasy with such a possibility. Why – and how – did the universe generate the potential for such an improbable precision within itself?

However, not everyone was coming to the conclusion that the universe actually was flat. Neta Bahcall tried to weigh the universe through galaxy clustering data, and in the process, calculated an omega value of 0.2 ... indicating that the tendency toward continued expansion appeared to be more powerful than a gravitational tendency toward collapse.

Ruth Daly worked with the data from radio galaxies and came up with a different value for omega. The value she calculated was 0.1 ... indicating that the universe was undergoing a form of continued expansion that was even greater than the value determined by Neta Bahcall.

In the 1990s members of the COBE (COsmic Background Explorer) project made a number of announcements concerning measurements of microwave radiation in the cosmos. Their data indicated the universe was flat ... that there was a degree of homogeneity present in that radiation involving deviations of just one part in 100,000.

But, what if Cosmic Microwave Background Radiation were due to some sort of on-gong process of thermal equilibrium that had arisen through ambient, cosmic dynamics and, therefore, did not necessarily represent remnants of the Big Bang? Wouldn't one expect the microwave background to be relatively homogeneous if it was at thermal equilibrium, and, consequently, isn't it possible that the universe might have been flat quite independently of considerations involving Big Bang expansion?

Furthermore, the aspect of omega that involves expansion is predicated on interpretive frameworks that are rooted in the belief that redshifts in the wavelength of light means such light is coming from an object exhibiting recessional velocity – that is, the source of light is receding from an observer. Nonetheless, in Chapters 2 ('The Meaning of Red'), 3 ('Noise'), and 5 ('Matters of Gravity'), possibilities were introduced indicating that the wavelengths of light might also be redshifted through the dynamics of new galaxy formation (Halton Arp), tired light, and the force of gravity.

If there had been no Big Bang, then, the notion of omega is problematic on several grounds. First, if there were no Big Bang, then, it doesn't necessarily follow that the universe has been, and is, expanding, and, secondly, if the presence of redshifted wavelengths does not necessarily indicate the presence of recessional velocity, then one cannot automatically assume that redshifted wavelengths in a light source means that the universe is expanding.

If expansion did not, and is not, taking place, then, the meaning of omega becomes immersed in arbitrary considerations. If the meaning of omega is problematic and doesn't measure what it purports to, then how is one to interpret the aforementioned calculations of Neta Bahcall, Ruth Daly, and others involving omega?

Calculations of omega presuppose that expansion has occurred and is continuing to occur. The dynamics of expansion presuppose that – quite independently of the Big Bang – redshifted wavelengths signify recessional velocities and/or increased distance.

If redshifted wavelengths can be generated through means other than recessional velocities or distance, then, the primary evidence for expansion becomes unreliable. If one wishes to retain the connection between redshifted wavelengths and an expanding universe, then, one needs to demonstrate that such redshifts are not a function of: (1) the dynamics associated with the birth of galaxies involving quasars; (2) the tired light phenomenon; and, (3) the impact of the force of gravity, or one needs to demonstrate that one can meaningfully distinguish between the contributions of the three factors noted above and the contribution of recessional velocities or the stretching of space when it comes to the issue of expansion.

Moreover, one cannot automatically assume that if the universe is not expanding that the ultimate fate of the universe would end in gravitational collapse. For example, if various forces – such as rotational spin, movement through space, gravitational attraction, and so on -- all shape what happens to a given cosmic body, then when one factors in all of those interactional dynamics, the system might, or might not, be in a state of stable equilibrium, and, therefore, be able to resist the tendency toward gravitational collapse.

To be sure, there are regions in the Milky Way galaxy – as well as among galaxies – where the prevailing set of interacting forces are unstable, and, as a result, those cosmic entities, eventually, will be drawn into one another. However, no one knows to what degree the universe as a whole might be in a condition of relative stability or instability with respect to the dynamics of interacting and, possibly, countervailing forces.

In addition, no one knows to what extent dark matter might serve as a stabilizing force in the universe. This is especially the case if there is more than one form of gravitational attraction, and, if those forms of gravity mitigate or alter, in some fashion, the impact of one another's presence … perhaps the interaction – or lack thereof – between dark matter and Standard Model matter serves to buffer the universe in various ways from undergoing total gravitational collapse.

If the foregoing issues are not sufficiently problematic, there is a related issue that introduces an additional problem. It concerns what is referred to – facetiously -- as the "axis of evil".

More specifically, although, for the most part, the Wilkinson Microwave Anisotropy Probe (WMAP) indicated that the Cosmic Microwave Background was incredibly homogeneous, there also were asymmetries present in the data. Some of those asymmetries were consistent with the Standard Model of the Universe, while some of those asymmetries could be the tip of a very problematic iceberg.

The useful asymmetries in the WMAP data were the ones (measuring ten millionths of a degree or less) that researchers believed marked the presence of the cosmic seeds that, over time, would develop into the large-scale structures of the universe. The problematic asymmetries involved variations in temperature that were unexpectedly large and were aligned in a manner that did not appear to be random (as had been expected) and, moreover, did not coincide with any known structure in the universe today.

Researchers have drawn a curving line through a visual representation of the microwave data (this representation is known as a power spectrum) that designates the areas of temperature variation and alignment anomalies. Kate Land and Joao Magueijo took a phrase – "the axis of evil" that had been uttered by George W. Bush in relation to a number of countries -- and, with tongue firmly planted in cheek, they applied the phrase to the line that ran through the areas of anomalous temperature variations and peculiar alignments in the power spectrum for the CMB data.

Independently, the Planck space mission that is operated through the European Space Agency verified the findings of WMAP. The unexpectedly large temperature differential and peculiar alignment in the areas of interest were not just artifacts of WMAP ... the data appeared to be reliably substantial in some sense.

What did the anomalies mean? Did they possess some unknown significance involving new physics, or were they merely localized manifestations of random events that were apropos to nothing?

According to one parsing of the WMAP/Planck data, one should expect that temperature variations might be larger or smaller from place to place in the Cosmic Microwave Background Radiation data. Chance events – supposedly -- will produce those kinds of variations.

Unfortunately, researchers really don't have a reliable measure for determining the sort of event that can be written off as being merely random in nature. Although chance events might account for the anomalous data of the Axis of Evil, why should one have an expectation that chance dynamics alone should be capable of producing the anomalous asymmetries in the WMAP/Planck data?

Do the temperature differentials of the universe conform to a normal distribution or some other kind of distribution, and, furthermore, irrespective of whatever kind of distribution those temperature differentials reflect, how does one determine that such unexpected and anomalous events are to be expected on the basis of chance alone? What model of chance are we talking about?

Can one even assume that the universe operates in a random fashion? Isn't this imposing a certain set of methodological filters on the data rather than letting the data speak for itself?

Randomness is not a neutral standard against which one can measure the events of the universe. The idea of randomness has ontological and metaphysical biases built into it.

How does one distinguish between events that are supposedly random in nature and events that manifest themselves in the same way as the allegedly random events but might not be random in nature? To be able to distinguish between the foregoing events, one has to have a test of significance of some kind, and irrespective of how extensively such tests might be applied and accepted, all those tests tend to establish arbitrary cut off points for differentiating between the random and the non-random.

According to some individuals, the anomalous data generated through WMAP and the Planck space mission can be dismissed because if one were able to observe the universe from the vantage point of a sufficiently large enough scale, then – supposedly – such anomalies would be observed to fit into an overall pattern of cosmic homogeneity mixed with the right amount of anisotropy to form the universe we see today. If this were the case, then, the WMAP/Planck findings would have no overarching significance.

To be sure, models of cosmic variance are used to interpret the WMAP/Planck data. However, even by the standards of those models, the asymmetries present in the WMAP/Planck data are unexpected and, possibly, problematic.

More importantly, we do not possess the sort of aforementioned vantage point that would be provided by placing such data in the context of a larger scale perspective. Consequently, the meaning of the anomalous asymmetries in the WMAP/Planck data cannot automatically be considered as expressions of chance happenings, nor can they automatically be dismissed as having been taken out of the sort of sufficiently large-scale context that might be able to show how such anomalies are merely normal expressions of cosmic variance.

Various individuals have introduced a number of tentative explanations in an attempt to account for the so-called 'Axis of Evil' that seems to be present in the WMAP and Planck space project data. However, to date, none of those offerings have gained much, if any, traction as a viable and reliable way to handle the data involving temperature variation and alignment in conjunction with the significant anomalies that appear to exist in Cosmic Microwave Background Radiation.

Among other things, the foregoing "Axis of Evil" data entails potential problems for the issue of expansion. At the very least, if that data represents something other than being an expression of normal, random cosmic variation, such data indicates that to whatever extent the universe expanded, then that process of expansion deviates, at least in part, from the version of things laid out by the Standard Model of cosmology

The expansion side of things is not the only source of problems for omega. The mass aspect of the omega ratio entails possible problems as well.

For example, astronomers have been studying clusters that contain thousands of galaxies. Such clusters provide an important source of data for exploring both the nature and extent of dark matter, as well as help to provide a means of piecing together an understanding of how the universe might have developed over time.

Researchers have compared sets of data from the Planck space project, the XMM-Newton Satellite, and the Sloan Digitalized Sky Survey. Such sets of data involve, respectively, millimeter wavelength, X-ray, and optical imagery probes.

Unfortunately, the foregoing three probes seem to provide significantly different estimates concerning the masses that are contained in the clusters being studied. Oftentimes, apparently, while any two of the aforementioned three methods might yield results that agree with one another, as soon as one tries to add results from the third method, inconsistencies seem to arise.

The mass of something should not depend on the method one uses to measure it. Nonetheless, millimeter wavelength, X-ray, and optical modes of analysis seem to bring about just such an unwelcome situation, and what implications, if any, the foregoing problem has for omega is unclear at the present time.

The inflationary theory of the universe that, initially, had been introduced by Alan Guth – and, since, has been amended somewhat -- predicted that the universe was flat. In other words, the inflationary theory maintained that the dynamic between the expansionary cosmological forces inherited from the Big Bang working in conjunction with the total mass of the universe, when taken together, suggested that the universe would neither collapse due to gravitational attraction nor continue to expand ... i.e., omega was 1.

However, empirical evidence from a variety of sources indicated that the universe seemed to contain 20% of what was needed to prevent the universe from continuing to expand. That is, based on a variety of data, omega seemed to be 0.20.

If the foregoing data is accurate, then, 80% of what is needed to yield an omega of 1 is missing. Alternatively, maybe the available evidence is hinting that the universe is not flat and, therefore, the inflationary theory of the universe – which made such a prediction – is wrong ... either partially or entirely.

For a number of decades, the empirical advantage arrow swung back and forth between theoretical perspectives claiming that omega was 1 and conceptual positions indicating that omega was less than 1. Some theorists even began to entertain the possibility that Einstein's discredited notion of $\Lambda$ – the cosmological constant -- might be able to help resolve the omega problem.

In the 1990s, two research groups – the High-Z Supernova Search Team, founded in 1994 by Brian Schmidt and Nicholas Suntzeff, and SCP (Supernova Cosmology Project), headed by Saul Perlmutter – began investigating Type 1a Supernovae in an attempt to determine, among other things, what implications, if any, such cosmic objects carried for the omega issue. Type 1-a Supernovae are distinguished from other types of supernovae through a spectral property that exhibits a strongly ionized silicon absorption line.

Type 1-a Supernovae are of particular interest because they constitute a class of entities that for the most part tend to display a constant brightness or luminosity when they explode ... although one might note there are some exceptions among Type 1a Supernovae that show a non-standard form of luminosity. The property of constant brightness permits such supernovae to be used as a "standard candle" for determining: (a) Cosmic distances (based on how faint the supernova appears) and (b) the degree to which space is stretched (which is indicated through the redshifts that are present in the wavelengths of the light given off by the supernova).

Calculations had been done in the 1990s that provided a value for the rate at which the university supposedly was expanding. The two, aforementioned research groups were particularly interested in comparing the rate of expansion taking place in the modern universe relative to the rate of expansion at various past junctures of cosmic history.

By determining how much the wavelengths of light from the supernovae were stretching or being redshifted over time, one might have a means for comparing the rates of expansion throughout that temporal period. If the redshift at a later point in time were more stretched out (i.e., more toward the red end of the spectrum) than was the redshift for an earlier point in time, then this would considered to be an indication that the universe – or the space therein -- had been expanding between those two points in time.

Both groups were expecting to uncover data that confirmed the conventional wisdom of the day – namely, that the expansion of the university was slowing down due to the presence of sufficient quantities of matter (and associated gravitational attraction) that could serve as a break on the rate of expansion. However, quite unexpectedly, both groups recorded data that, upon analysis, not only suggested that the rate of expansion in the universe was speeding up, but, as well, there did not seem to be enough matter available to give expression to a flat universe.

More specifically, the supernovae being studied by the two groups appeared to be dimmer than expected. Given the tendency of Type 1-a Supernovae to exhibit a standard brightness, the unexpected dimness of the supernovae being observed seemed to indicate that they were further away than anticipated, and, if this were true, then, conceivably, the rate of expansion for the universe might be accelerating and not decelerating.

However, if the universe were accelerating rather than – as conventional wisdom had maintained -- decelerating, then what was subsidizing the expansion? The answer seemed to be something called dark energy, and the "dark' part of the term alluded to the unknown nature of that energy.

When the foregoing results were released in 1998, there was considerable resistance to accepting them. The research findings were so unexpected that some astronomers felt the conclusions that were based on those results were premature ... such scientists felt that not only did the data need further vetting but additional data should be gathered as well.

Eventually Adam Riess and Brian Schmidt, along with Saul Perlmutter, would receive a Nobel Prize in 2011 for their work involving Type 1-a Supernovae. However, prior to releasing those results, Riess was hesitant to mention his findings to anyone because of the surprising nature of the discovery, while Schmidt felt fairly confident that many scientists would be inclined to reject the research due to the way it undermined conventional wisdom in such an unexpected manner ... for example, that the rate of universal expansion was speeding up, not slowing down and that such a finding implied the existence of an, heretofore, unanticipated form of energy.

Indeed, various scientists were skeptical of the Type 1-a Supernovae data. Some of these skeptics entertained the possibility that, maybe, the reason why some of the supernovae appeared to be dimmer than anticipated was not due to the factor of increased distance but, instead, was because there might have been more cosmic dust floating about in the universe than previously was believed to have been the case and that such dust might have made various supernovae appear to be dim.

Other skeptics considered the possibility that the unanticipated dimness of the original observations involving supernovae might be due to other factors. For instance, what if some supernovae explosions were less energetic than others, and, as a result, were less bright?

In 2000, the research team headed by Saul Perlmutter, used observations derived through the Hubble Telescope to study 12 more Type 1-a Supernovae. They found no more dust within galaxies than previously had been thought to be the case.

Even more emphatically, in 2002, Adam Riess assembled a team that used the Hubble Telescope to investigate a further 25 additional Type 1-a supernovae. The data obtained and analyzed by this research team appeared to indicate that possibilities such as dust and less energetic explosions could not account for the new data as well as a model that described a universe undergoing an increased rate of expansion due to the presence of dark energy.

Since the advent of the research on Type 1-a Supernovae, astrophysicists have gradually adopted the following understanding concerning the cosmology of the universe. First, the universe – under the influence of the Big Bang and inflationary theory – expanded fairly rapidly.

However, in the beginning, the universe was also fairly small and dense. Consequently, over time, the presence of matter began to decelerate the expansion of the universe.

During the Big Bang as well as the period of inflation, but before the expansion of the universe started to slow down, another kind of energy – namely, dark energy – had been running in the background. Initially, the effect of that energy was weak relative to the strength of gravitational attraction, but as the universe continued to expand – albeit at a slower rate – matter began to become separated by increasing distances, and, as a result, the effect of dark energy began to play an increasingly important role in shaping he universe.

In short, as cosmic distances increased, the force of gravity weakened Simultaneously, as the force of gravity began to diminish, the effect of dark energy became more pronounced, and when this occurred, the rate of universal expansion began to pick up steam.

Both of the research groups (the High-Z and SCP teams) that had been working on the Type 1-a Supernovae issue published their findings and conclusions in October 1997. Each of the teams maintained there was not enough matter in the universe to generate an omega of 1 – that is, a flat universe.

There seemed to be two possibilities. (1) The cosmological constant, $\Lambda$, was 0, and omega was low, so the universe would be open and continue to expand. (2) Omega was low, while the cosmological constant of general relativity, $\Lambda$, was non-0, and, as a result, expansion would slow down but the universe would continue to expand.

If option (2) turned out to be true, then $\Lambda$ would rise from the cemetery of defunct ideas. Einstein's biggest blunder would become an important part of an equation that was better than he had come to believe following the publication of data indicating that the universe might be expanding ... data that led Einstein to reject the notion of a steady state universe, and, therefore, $\Lambda$.

Most mainstream astrophysicists were proponents of a modified version of the inflationary theory that had been proposed by Alan Guth in the late 1970s. The theory of inflation required a flat universe – that is, an omega of 1.

As noted earlier, both of the Type 1-a Supernovae research teams indicated there was not enough matter in the universe to bring about a flat universe by virtue of matter alone. However, if the cosmological constant were to contribute the right amount of energy density to the universe, then, one still might end up with a flat, or nearly flat, universe.

While checking over the Type 1-a Supernovae research in order to make sure it didn't contain any errors with respect to data involving, among other things, dust, Adam Riess generated calculations that appeared to describe the universe as consisting of negative amounts of matter. Those results didn't make sense unless they were interpreted as giving expression to a positive value for $\Lambda$ ... something that most astrophysicists were reluctant to do in light of the history of controversy that had surrounded the cosmological constant ever since Einstein acknowledged his mistake.

Riess forwarded his results to his team leader, Brian Schmidt (who along with Riess would win a Nobel Prize for their Type 1-a Supernovae work). Schmidt would go through the data independently, make his own calculations, and, then, compare his results with the ones that Riess had sent him.

When he completed his calculations, Schmidt agreed with Riess. The Type 1-a Supernovae data and calculations pointed to the existence of a positive value for Lambda, Λ, and this could be asserted with a statistical level of confidence reaching 99.7.

A positive Λ might be playing several roles. On the one hand, its energy density would help to slow down the expansion of the universe, while, supposedly, simultaneously also giving expression to a force that helped the universe to expand.

Between the properties of being (possibly) a force of expansion as well as being a source of energy density that slowed the universe down, dark energy -- to which lambda, at least in part, contributed -- appeared to account for 70%, or more, of the "stuff" of the universe. Dark matter made up somewhere between 23-26 % of the matter of the universe, while Standard Model matter represented about 4% of the universe's composition.

The two Type 1-a Supernovae research teams had initially set out to establish the extent to which the universe might be slowing down (due to gravitational attraction) in relation to the expansive impetus that had been given by the Big Bang. Several years later they all arrived at the same conclusion: Not only wasn't expansion slowing down, but, even more astonishingly, the rate of expansion actually appeared to be speeding up, and this seemed to be due to some sort of dynamic involving dark energy, and, at least in part, Einstein's blunder seemed to play a role in all of this.

Although – as previously indicated -- a Saul Perlmutter led research team in 2000, as well as a Adam Riess led team in 2002, both provided empirical data indicating that dust could be ruled out as a possible reason for why the Type 1-a Supernovae being observed were dimmer than anticipated, there is also data that has been uncovered after the aforementioned research had been published which raises the possibility that the act of ruling out dust as a possible source of dimness in the Type 1-a Supernovae might have been premature to some extent. More specifically, the Planck space telescope has gathered data suggesting there might be much more dust in the universe than originally had been thought to be the case.

Leaving aside the dust issue for now (this topic will resurface shortly), there are a few other problems that pose a serious challenge to the Type 1-a Supernovae research ... especially in relation to the way in which that research has been interpreted. Several of these problems were touched upon earlier when discussing the idea of an early, inflationary universe within the context of a Big Bang scenario.

To begin with, there is the issue of whether, or not, space is capable of being inflated. No one knows what space is, and no one understands the nature of the dynamics, if any, through which space might be inflated.

Transitions in the redshifts of light coming from Type 1-a Supernovae are being used as the primary evidence in support of the idea that space appears to be inflatable. Originally, back in the late-1920s, when Hubble was introducing his ideas about the possibility of an expanding universe, redshifts were interpreted to mean that the stars and galaxies displaying that sort of redshifted light were moving away from observers on Earth, and, thus, those spectral changes were understood to give expression to a "recessional velocity" for cosmic objects exhibiting redshifts.

However, if the redshifts associated with Type 1-a Supernovae were interpreted to mean "recessional velocities", then some of those stellar objects would be traveling at improbable, if not impossible, velocities relative to the speed of light. Consequently, the causal responsibility for the observed redshifts was transferred to the manner in which the expansion of space appeared to stretch out the wavelength of light and, in the process, pull wavelengths toward the red end of the spectrum.

Although grasping the general idea of expanding space is easy enough to do, many unknowns begin to rush in when one tries to take a closer, more concrete look at the dynamics associated with the notion of expanding space. For example, what is the nature of the energy that inflates space, and what is the structural nature of space that renders it sensitive to the presence of that sort of energy, and what are the details of the interaction between space and energy that results in expansion, and how does energy become transformed into more space?

Does space become thinned in some fashion as dark energy stretches it out or is more space being generated? If the former is the case, then, what is the nature of space that permits it to be thinned out, and if the latter is the case, then how does dark energy get converted into additional space?

Moreover, proponents of the idea of expanding space believe that as much as 70%, or so, of the total amount of mass and energy that exists in the universe is tangled up, somehow, with space. Since energy also has a mass equivalency, and, therefore, a gravitational potential, then, presumably the reason why the presence of such massive amounts of energy haven't induced the universe to move toward gravitational collapse over time is because dark energy's potential for expansion is greater than its potential as a source of gravity.

Dark energy's internal dynamic involving its expansive and gravitational properties must be an interesting one. Among other things, one might wonder about how energy density generates gravitational force ... that is, to borrow a word from quantum gravity theory, how does energy density give rise to gravitons.

Even if one were inclined to accept the idea of expanding space, one isn't necessarily required to argue that the source of the inflating energy must reside, somehow, in space itself. When the "space" within a balloon is increased, it is not the "space", per se, within the balloon that expands, but, rather, "space" is introduced from outside of the balloon, and the energy through which that "space" is introduced is external to the inside of the balloon.

Conceivably, if space actually is capable of expanding, the energy that subsidizes such expansion could be extra-dimensional. Maybe space is a semi-permeable dimension that allows certain other dimensions to interact with it (e.g., an inflationary dimension), just as the inhabitants of Flatland were able to interact, within limits, with a three-dimensional being ... and vice versa.

If the foregoing possibility correctly describes the nature of expanding space, then the energy that subsidizes the expansion of space could come from some dimension other than space. To be sure, a certain amount of energy might be introduced into space during the inter-dimensional transaction, but the majority of the energy involved in such a process would not necessarily have to be inherent in the dimension of space.

Introducing "space" into a balloon doesn't violate any laws of conservation because it is the total system that must be taken into consideration in relation to those laws. Similarly, inter-dimensional transactions involving energy and space need not violate any laws of conservation since it is the total system of the universe that must be considered irrespective of the number of dimensions that are entailed therein

Notwithstanding the foregoing possibilities, there is a fair amount of tangible evidence to indicate that redshifts in wavelength do not necessarily involve either the stretching of space or recessional velocities. Earlier in this book ('Chapter 2: The Meaning of Red'), some of the research of Halton Arp and Margaret Burbidge involving quasars was explored, and during that discussion, a paper was mentioned that had been presented in 2004 by Margaret Burbidge, her husband, Geoffrey, Halton Arp and several other individuals.

The foregoing article was about the relationship between a highly redshifted quasar and a low redshifted galaxy – namely, NGC7319. The highly redshifted quasar was located in front of the low redshifted galaxy, thereby indicating that the disparity in redshift had nothing to do with recessional velocities or the stretching of space.

The wavelengths of the light coming from the quasar give expression to what is referred to as "intrinsic redshift". In other words, there is something intrinsic to the dynamics of the quasar that is generating such high redshift measurements, and those redshifts are not a function of distance, recessional velocities, or the stretching of space.

In 'Chapter 3: Noise', as well as in 'Chapter 5: Matters of Gravity', several topics were explored that focused on, respectively, the way in which light can be redshifted through the dynamics of the journey that light undergoes while traveling through millions of light years of space that is far from empty, and, as well, how light is also redshifted through the presence of gravity – and, one should keep in mind, that if 70% of the "stuff" of the universe is dark energy, then, light is constantly traveling through an energy density that has a mass, and, therefore, a gravitational equivalency. While one can make estimates about the extent to which light might be redshifted through such events, those estimates are often rather arbitrary and, consequently, are much more akin to guesses than reliable calculations.

Some people guess that the impact of such factors on redshift will be negligible, while other individuals guess that the impact of those factors on redshift will be substantial. However, there is no real hard data on either side of the issue that suggests one kind of guess is more accurate than some other kind of guess is.

The issue of intrinsic redshift, on the other hand, is an entirely different matter. Providing hard data that a quasar closer to Earth is more highly redshifted than a more distant, low redshifted galaxy strongly indicates that, at least in some instances, redshift appears not to be a function of either recessional velocity or the stretching of space.

Although various individuals have taken issue with the Burbidge[2]/Arp 2004 discovery, the 2004 data have not been overturned by any subsequent empirical findings. Nevertheless, the foregoing concrete fact continues to be ignored by proponents of the Standard Model of Cosmology ... a model that is deeply entangled in the belief that redshift necessarily means either recessional velocity, the stretching of space, and/or cosmic distance.

In 1984, Michael Turner, Lawrence Krauss, and Gary Steigman wrote a paper bearing the title: "Flatness of the Universe: Reconciling Theoretical Prejudices with Observational Data". The paper was an attempt to show that a theoretical prejudice involving the idea of a flat universe might be consistent with certain observations that represented a possible challenge to the prejudice to which they were alluding in the title of their paper.

Irrespective of whether, or not, the foregoing authors were successful in their attempt at reconciliation, there are two points to note in conjunction with the aforementioned article that are relevant to the present discussion. Firstly, although one doesn't often see such clear admissions in the literature, the title of the foregoing article seems to indicate that "theoretical prejudices" might play prominent roles in science ... even if the authors might have had their tongues firmly planted in their cheeks when they conceived of the title for their article.

Secondly – and this is a direct result of embracing such theoretical prejudices -- there is a natural inclination among many scientists to try to preserve the appearance of viability in relation to various, mainstream theories (e.g., an inflationary theory that predicts a flat universe) by explaining away (i.e., reconciling) empirical data in a manner that permits such theories to continue on as if they were the best way to account for the data ... even as other evidence indicates that such an attempt at reconciliation might be ill-conceived.

Adam Riess, one of three Nobel Prize recipients for the cosmological discovery involving Type 1-a Supernovae, once mentioned that the measurements underlying the Type 1-a research are complicated, and, consequently, there are many points along such a chain of measurement that might entail an error of some kind. He claims – and I believe him to be very sincere and honest in his contention, as well as a highly competent scientist – that he went back over the chain of measurement process and not only were there no sources of error uncovered during his critical inquiry into the viability of the Type 1-a data, but, in addition, Riess felt that upon completing his check of the data, there didn't appear to be any way to interpret the data other than to conclude that the universe was expanding and that such an expansion was being fueled by the presence of dark energy.

The phenomenon of intrinsic redshift that had been advanced in the aforementioned Burbidge[2]/Arp 2004 paper appears to constitute an alternative way of engaging the Type 1-a Supernovae data. More specifically, isn't it possible – perhaps even reasonable – to argue that the sorts of redshifts that were exhibited by the Type 1-a Supernovae are a function of some combination of intrinsic redshift, tired light, and the influence of gravity on the wavelengths of light coming from those supernovae rather than being an indication that the rate at which space expanded was increasing and that this was due to the presence of dark energy?

Although Halton Arp believes that intrinsic redshift has something to do with the dynamics underlying the birth of new galaxies, neither Arp, Margaret Burbidge, nor Geoffrey Burbidge know – in precise terms -- what causes intrinsic redshift. They only have found evidence indicating that the phenomenon seems to be real and, as a result, one might have to look to some other kind of dynamic besides recessional velocity and/or the stretching of space to account for the phenomenon of intrinsic redshifts.

Of course, the Type 1-a Supernovae data were released six years before the aforementioned Burbidge[2]/Arp paper. However, Halton Arp had been presenting similar kinds of evidence (although, possibly, more circumstantial in nature) for quite a few years before the Type 1-a research was undertaken ... evidence that pointed in the direction of a phenomenon – i.e., intrinsic redshift – that was not tied to the usual ways of interpreting the meaning of redshifts (i.e., as indicating distance, recessional velocity, and/or the stretching of space).

Most (but not all) of the movers and shakers in the realm of astronomy responded to Arp's research by denying him access to the very instruments (i.e., telescopes) that were necessary to further explore the parameters of the intrinsic redshift phenomenon (or to uncover data indicating that the idea was spurious). In other words, theoretical prejudices seemed to be intent on making the world of astronomy safe for only certain kinds of theories (e.g., inflation, the Big Bang, and an expanding universe).

Perhaps the reason why Adam Riess didn't find any source of error during his check of the Type 1-a Supernovae data is because he – like so many of his colleagues -- was unwilling to critically question the soundness of an underlying theoretical prejudice which held that redshift necessarily must be tied to recessional velocity, distance, and/or the stretching of space. In any event, claiming that the Type 1-a Supernovae data points inexorably in the direction of expanding space and dark energy appears to be somewhat premature.

Perlmutter, Riess, and Schmidt – the Nobel Prize winners for the Type 1-a Supernovae research – did not demonstrate that space possesses a nature that is capable of being expanded, nor did they demonstrate how some sort of energy could be translated into the expansion of space (through being thinned or through the generation of additional space). Dark energy and an expanding universe were a hermeneutical rendering of data that could be interpreted in at least one other way.

For example, evidence did exist (through the work of Arp, the Burbidges and others) long before the Type 1-a Supernovae research got underway indicating there is a form of redshift that does not appear to be a function of recessional velocities, distance, or the stretching of space. Consequently, why should anyone jump to the conclusion that the redshifts present in the Type 1-a Supernovae data necessarily meant that space is being stretched?

This is something of a mystery until one acknowledges the possibility that theoretical prejudices might have prevented mainstream astrophysicists from seriously entertaining alternatives to the sort of framework concerning redshifts that conceptually has filtered nearly everything done by astronomers and astrophysicists for more than eight decades. Apparently, it is easier for such individuals to claim, without any real evidence, that space expands than it is for them to acknowledge – despite the presence of considerable evidence -- that redshifts don't necessarily mean recessional velocities and/or the stretching of space.

Evidently, it is more scientifically appropriate to invent the existence of something called dark energy than it is to consider the possibility that there is no need to invent something of a mysterious nature. Conceivably, the issue of intrinsic redshift (along with the idea of tired light and the way in which gravity tends to stretch the wavelength of light in the direction of the red end of the spectrum) might preempt the need to do so.

Maybe space does expand. Perhaps dark energy does exist. Maybe, the presence of redshifts in Type 1-a Supernovae indicates that space is being thinned and/or generated.

However, until one can show, clearly and persuasively, the precise nature of the dynamics governing the manner in which dark energy causes space to expand, then the data derived from Type 1-a Supernovae research does not, on its own, appear to demonstrate that space stretches and that dark energy exists. This is not to say that the current interpretations of the Type 1-a Supernovae data might not be correct, but, rather, it is to say that quite a few more pieces of reliable evidence need to be forthcoming before such conclusions would appear to be truly warranted.

Notwithstanding the foregoing considerations, for many astrophysicists, the theoretical pull of a positive lambda (which is mixed in with the issue of dark energy) is extremely strong. Through a positive lambda, the idea of inflation can be saved, and through inflation, a number of problems (such as isotropy and homogeneity) that are entangled with the Big Bang scenario can be resolved.

A positive lambda has a strong upside, but it is not all milk and honey. The existence of a positive lambda entails a problem for particle physicists.

Particle physicists consider space to be an "active" participant in the universe. They do not believe that space is empty but, instead, they maintain that space gives expression to an array of virtual particle activity.

Consequently, for many physicists, to whatever extent lambda exists and contributes energy density to omega, lambda seems to be a property of space. Nonetheless, a distinction can – and, perhaps, should -- be made between, on the one hand, a given energy density that exists within a field that is considered to be vacuum-like, and, on the other hand, the nature of space in and of itself.

Whether one considers lambda to be a property of space or an expression of a given vacuum state, particle energies do interact with gravity. Hendrik Casmir had reflected on this latter issue in 1948 when he hypothesized that increases in vacuum energy should be measurable as one brought two conducting plates closer to one another.

The foregoing phenomenon is known as the Casmir effect. Empirical results have confirmed the reality of that phenomenon.

What significance does the issue of energy density have for lambda and the cosmological constant? Many individuals were treating lambda and the cosmological constant as being synonymous with one another.

Michael Turner, among others, wanted to put a conceptual speed bump, of sorts, in the way of those people who were automatically assuming that lambda and the cosmological constant were synonymous with one another. Therefore, he came up with the idea of "dark energy" as a way of alluding to a form of energy vacuum that would vary across space and time and that could be distinguished from a form of vacuum energy – namely, the cosmological constant -- that would remain constant across space and time.

Astronomers had estimated the positive energy density of space to be somewhere between: 0.6 to 0.7. Quantum and particle physicists had arrived at a slightly different estimate for the magnitude of the energy density of space ... $10^{120}$

If space had an energy density in the vicinity of a magnitude like $10^{120}$, then, among other things, the temperature of the cosmic background radiation would have cooled to below 3 degrees Kelvin in a tiny fraction of a second following the Big Bang. Perhaps even more importantly, there seems to be no reasonable or plausible way to reconcile the estimates of astronomers and particle physicists when it comes to the issue of predicting the amount of energy density that exists in space or to which space gives expression.

As has been pointed out elsewhere in the present book, as well as in *Volume II* of *Final Jeopardy*, perhaps one of the problems with the estimates of particle physicists in relation to the energy density of space involving virtual particles is that such particles – to whatever extent they exist – are far fewer in number (perhaps a gross understatement) than particle physicists suppose is the case. To be sure, the Casmir effect, along with other empirical results, indicate that a certain amount of energy density exists in a vacuum state, but there is no current way to precisely measure just what magnitude of energy density is actually present.

If the foregoing notion is correct, then there is something substantially wrong with the way particle physicists go about making estimates concerning the energy density of space or the energy density that is present when a vacuum condition exists. And, perhaps, the 'something that is substantially wrong' with such estimates is rooted in the way particle physicists understand the idea of virtual particles ... what they are, or how they work, or what causes them, or how many of them there are at any given point in time and space.

When Riess presented his aforementioned follow-up dust studies in 2001 and 2002 involving Type 1-a Supernovae research, he put together some visual representations that displayed a point of transition in which the expansion of the universe appeared to go from a stage of deceleration to one of acceleration. However, if Arp, Burbidge[2], and others are correct concerning their ideas about how redshift does not necessarily mean the presence of either recessional velocity or the stretching of space, then, what Riess's charts showed might, indeed, have been a transition of some sort, but that visual material did not necessarily depict a transition that went from a decelerating rate of expansion to an accelerating rate of expansion for the universe.

Let's take a closer look at the matter. First, redshift and dimness (the two themes that are at the heart of Type 1-a Supernovae research) are – potentially -- separate issues.

More specifically, the redshifts cited in the foregoing research are considered to demonstrate that the variable degrees of dimness found among Type 1-a Supernovae is a function of the stretching of space. However, if such redshifts were, instead, shown to be due to some other set of conditions (such as intrinsic redshift, tired light, presence of gravity, recessional velocity, etc.), then, an obvious question to ask is the following one: If space is not expanding, then, why do cosmic objects that usually exhibit a standard brightness (i.e., Type 1-a Supernovae) exhibit dimness under certain conditions?

In other words, if redshift does not necessarily mean that space has been stretched through expansion, then the issue of dimness still has to be explained. As such, the problem of dimness is an issue that exists independently of the redshift problem.

A possible, alternative way to account for the observed differences of brightness in Type 1-a Supernovae involves dust. Riess employed a technique – MLCS (multicolor light-curve shapes) developed in his doctoral dissertation -- to correct for the presence of dust in relation to redshift.

Let's leave aside the fact that more than a decade after Riess had eliminated dust as a possible explanation for the dimness of Type 1-a Supernovae, the Planck space project uncovered evidence indicating that there might be more dust in space than originally had been thought to be present. Instead, let's assume that galactic dust – at least when considered in the usual sense -- does not account for the dimness observed among various Type 1-a Supernovae. If not dust, then what?

Although the stretching of space is the explanation that was given by the Type 1-a Supernovae research teams to account for the dimness issue, there doesn't appear to be anything in their research which demonstrates that the redshift observed in conjunction with Type 1-a Supernovae could not be due to the presence of some combination of: Intrinsic redshift dynamics (whatever that turns out to be); tired light; gravitational fields; and/or recessional velocities. In other words, although the role that dust plays in affecting redshift was examined, none of the other possible sources that might affect redshift appeared to have been examined with the same care and consideration as was accorded to dust.

For example, as far as the significance of redshift is concerned, how does one distinguish between, on the one hand, any recessional velocity that might be displayed by the light from a supernova and, on the other hand, the stretching of space that might have occurred between that object and the observer? Obviously, if a given redshift indicates that a supernova would have to be traveling with a velocity that is near, at, or in excess of the speed of light, then, one would have to discount recessional velocity as being the sole cause of such a redshift, but, there is nothing that prevents one from considering the possibility that recessional velocity might make up some portion of the observed redshift.

Various cosmic objects travel toward one another (as in the case of the Andromeda and Milky Way galaxies) and, in the process, their respective lights are blue shifted. Similarly, various cosmic objects travel away from one another, and, therefore, those objects will not only display a recessional velocity, but, as well, will exhibit a redshift in the wavelength of their light with respect to one another that will be a function of the recessional velocities of both objects.

What sort of a contribution such recessional velocities make to the degree of redshift will vary from situation to situation. This contribution might not be great, but it does have to be taken into consideration.

In other words, even if such a source of redshift is not significant, nonetheless, it plays a role when trying to account for just how much space is expanding (assuming that it is). If some portion of redshift is due to recessional velocity, then, to that extent, space is not being stretched (if this is what is taking place).

After taking recessional velocity into account, one has to consider the other sources that might contribute to the redshift displayed by various Type 1-a Supernovae. For instance, to what extent do: Intrinsic redshift, tired light, and gravitational influences contribute to the degree of redshift that is observed in conjunction with Type 1-a Supernovae?

All three of the foregoing possible sources of redshifting are not well understood – if even acknowledged -- by mainstream astrophysicists. Consequently, assigning some sort of value to the contribution made by such sources to redshift tends to be a rather arbitrary process.

At the very least, the ignorance surrounding the contributions that such sources might make to redshift should cause one to adopt a certain amount of caution when it comes to interpreting the meaning of redshift in conjunction with Type 1-a Supernovae. If nothing else, the degree to which the foregoing sorts of sources contribute to Type 1-a Supernovae redshifts will have to be subtracted from the extent to which space is believed to expand (assuming that it does), and since the magnitude of the former contribution is unknown and not necessarily negligible, one is not really in any position to determine whether, or not, space expands, and if it does expand, to what extent this occurs.

If – at least conceivably, if not somewhat plausibly – the redshift associated with Type 1-a Supernovae is not necessarily an indication that space expands, then, what might be causing varying degrees of dimness associated with such supernovae? There are a number of possibilities that bubble to the surface.

Type 1-a Supernovae tend to occur in binary systems in which at least one of the two stellar objects is a white dwarf. White dwarfs have completed their fusion lifecycle for the most part (occasionally, further fusion reactions are possible under the right circumstances).

The peak luminosity of such supernovae tends to be the same from one SN event to the next due, in large part, to the uniform nature of their masses. The relationship between peak luminosity and uniform mass permits Type 1-a Supernovae to serve as standard candles for measuring distances ... a relationship that figures very prominently in the research that led to the idea that space is expanding and that dark energy is fueling such expansion.

In addition to mass, certain kinds of atoms appear to play a role in the uniformity of the brightness displayed by most Type 1-a Supernovae. For instance, the amount of Nickel-56 produced seems to be related to the uniformity of the brightness that occurs at peak luminosity in relation to Type 1-a SN events.

Although, generally speaking, Type 1-a Supernovae do exhibit standard peak luminosity, this is not always the case. However, there is a way (known as the Phillips relationship) to correct for such differences in peak luminosity and, thereby, arrive at a value that affects the distance standard candle value of Type 1-a Supernovae by a factor of 7% ... obviously, such a value also carries some implications – relatively minor though they might be – for the issue of dimness.

A more substantial problem – potentially -- for the issue of dimness comes in the form of a relatively recent discovery. More specifically, apparently, not all Type 1-a Supernovae are the same, and one of the developments that has arisen out of the foregoing discovery is the fact that there is a version of Type 1-a Supernovae that is relatively rare today, but this was not always the case.

The empirical dust, so to speak, is still settling in relation to the foregoing issue. Therefore, it is uncertain to what extent the newly discovered information concerning the varieties of Type 1-a Supernovae will impact the issue of dimness.

Conceivably, the transition in brightness that was visually displayed in Riess's aforementioned presentation did not necessarily mark the crossover point between a cosmos that was decelerating in expansion to a universe that was exhibiting an increase in its rate of expansion. Perhaps, the transition point exhibited in the visual aides used by Riess marked a point when the newly discovered edition of a Type 1-a Supernovae was, for some unknown reason, becoming less common than previously had been the case, and this transition in relative numbers of the different kinds of editions of Type 1-a Supernovae might have affected the dimness data.

In addition to the foregoing consideration, there is another issue that might have some relevance here. More specifically, although

Adam Riess appeared to put forth a fairly rigorous set of reasons for discounting the idea that dust might have played a significant role with respect to either the issue of dimness or redshifts in relation to the Type 1-a Supernovae research, there is a variation on the dust theme that should be considered.

Such considerations might be especially important given that a lot of astronomers seem inclined to ignore the impact that the electromagnetic phenomena present in galactic and extragalactic plasmas might have on a variety of phenomena. This includes the issue of dimness.

For instance, in view of the fact that plasmas and associated electromagnetic forces affect weather on Earth in so many ways (e.g., lightning, tornadoes, hurricanes, and so on) one wonders whether, or not, the plasmas that are ubiquitous in outer space might help to generate and shape something akin to galactic and intergalactic weather patterns. For example, galactic and intergalactic dust can become ionized and, in the process, help lead to the formation of plasmas.

Moving particles of ionized dust can generate electromagnetic phenomena. Such phenomena might create galactic and intergalactic dust storms of varying kinds.

Like storms on Earth, as underlying galactic and extragalactic conditions change, the dust systems that exist in space could acquire different degrees of intensity as well as arise and be dispersed within different temporal frameworks. If the electromagnetic forces present in galactic and extragalactic plasmas created dust storms, the latter might occur anywhere along the line of sight between a Type 1-a Supernova.

As a result, the degree of dimness that is associated with such objects might be affected. This might not be the reason why unexpected dimness occurs in relation to all Type 1-a Supernovae, but it might be a reason in an unknown percentage of those supernovae, and if it were, the overall argument concerning the alleged nature of the relationship between redshift and dimness in conjunction with Type 1-a Supernovae might have to be re-calibrated to take that possibility into account.

The more distant a given supernovae is, the more likely it is there might have been galactic and extragalactic dust storms of the foregoing kind (assuming these actually occur) that could have impacted the appearance of brightness or dimness of the light that is being received by an observer. Maybe some Type 1-a Supernovae are dimmer than one would predict because their light has interacted with one, or more, galactic and extragalactic dust storms of varying intensity along the journey of such light to an observer.

One might be able to estimate the amount of dust that exists, on average, in the universe, and, then, remove the effects of such standard dust forms from one's calculations when discussing the dimness or redshifts of Type 1-a Supernovae. Nevertheless, how does one factor in galactic and extragalactic dust storms of varying sizes, intensities, and duration that might have arisen, and, then, dispersed as a function of the electromagnetic activity that takes place in the plasmas that occupy most of the universe?

Researchers like Adam Riess and Brian Schmidt use levels of statistical significance such as 99.7 to describe the confidence they have in their calculations and analyses. However, what do such levels of statistical significance actually mean?

They tend to mean that people who share the same: Biases, assumptions, models of probability, methods, and modes of calculations are likely to come to similar conclusions and have a similar degree of confidence in the process used to arrive at those conclusions. People who don't share such biases, assumptions, models, methods, and modes of calculation are not likely to have the same degree of confidence in those same conclusions.

Degrees of statistical significance tend to reflect one's degree of investment in the theoretical and methodological framework through which data is generated, filtered, processed, and interpreted. Degrees of statistical significance don't necessarily reflect the nature of reality

For instance, those individuals who – for what they believe to be good reasons – reject the notions of: The Big Bang, inflationary theory, dark matter, dark energy, the significance of cosmic background radiation, and the idea that redshifts necessarily signify distance, recessional velocity, or the expansion of space ... such individuals are likely to come to different conclusions concerning the meaning of the

Type 1-a Supernovae data than did Reiss, Schmidt, Perlmutter or those people (for example, the Nobel committee dealing with physics) who agree with the three aforementioned researchers. Those individuals who disagree with the mainstream perspective concerning the Type 1-a Supernovae data might even be willing to issue a level of statistical significance that is to be associated with their degree of confidence in the accuracy of their own way of doing things, but such a level of statistical significance is likely to be as arbitrary and system-biased as the aforementioned level of statistical significance (e.g., 99.7) that is associated with the idea that Type 1-a Supernovae research demonstrates that the rate at which the universe is expanding is increasing and this is due to the presence of dark energy.

Perlmutter, Riess, and Schmidt might be right in their assessment of the Type 1-a Supernovae data. While, if true, such a possibility certainly has some degree of relevance to the process of generating a response to the reality problem within the context of the Final Jeopardy challenge, what is far more important is to realize that there are many junctures along the chain of reasoning that leads from observations to the drawing of conclusions in relation to those observations, and all of these junctures are vulnerable to errors of one kind or another that can affect the quality and accuracy of one's conclusions.

This is the case with respect to Type 1-a Supernovae research. This is also the case with respect to life in general.

The possibilities that have been outlined in the last three to four pages might turn out not to be true. If so, and if the explanation for the unexpected dimness exhibited by Type 1-a Supernovae is not due to the stretching of space (an interpretation of the redshift that is associated with the light from those cosmic objects ... a redshift that might be explained through alternative possibilities), then, the dimness issue would remain something of a mystery.

Claiming that the unexpected dimness of Type 1-a Supernovae demonstrates that space is stretching and such stretching is caused by the presence of dark energy might provide a certain amount of satisfaction and conceptual closure. However, such closure carries a price that is potentially quite costly with respect to whether, or not, one actually has resolved the reality problem in relation to the nature of space and the issue of dark energy.

## Chapter 9: Branes for Hire

Most of the exploratory discussions that have taken place in previous chapters have either been critically directed toward one, or another, facet of the Big Bang scenario, or the latter theoretical framework has influentially occupied the conceptual horizons that have bordered and shaped the topics being critically explored. This is not because I am committed to some alternative, non-Big Bang cosmological account of the universe and, therefore, have been trying to poke holes in the Big Bang theory before presenting my own preferred framework.

As indicated in the Introduction for this book, I am indifferent to whether a true account of cosmology involves, or precludes, the possibility of a Big Bang. Empirical findings – past, present, and future – might demonstrate that every theoretical feature of the Big Bang framework has been substantiated or disproven, and neither possibility undermines my overall hermeneutical position vis-à-vis the reality problem.

This is not to say that my perspective is immune to facts. Rather, it is to say that whatever the cosmological nature of the universe turns out to be, those truths could be incorporated into the fabric of my understanding without appreciably altering or threatening the basic principles that form the underpinnings of such an understanding.

One of the foundational principles being alluded to in the foregoing paragraph is that whatever the truth turns out to be, one should acknowledge those truths and work with them. Tangled webs aren't created just through the process of deception, but, as well, tangled webs arise when one begins to enter into a condition of denial with respect to whatever the nature of truth involves.

Nonetheless, at the same time, determining what constitutes the truth in any given context is not always a straightforward matter. Although the Big Bang theory is certainly the reigning champion among mainstream astronomers and astrophysicists, that account has many moving parts that do not all function flawlessly and are not necessarily devoid of their share of unresolved problems … and some of those problems have been critically explored during the last several hundred-plus pages.

Scientifically speaking, over the last hundred years, or so, the Big Bang theory has had only one rival. That competitor is referred to as the Steady State theory but, for the most part, that model has been considered to be defunct since the issue of Cosmic Microwave Background Radiation came to prominence in the mid-to-late 1960s.

In 2007, however, two individuals – Paul J. Steinhardt and Neil Turok – released a book entitled: *Endless Universe: Beyond the Big Bang* that sought to advance their own perspective concerning the nature of cosmology. Since so much time has been spent during the preceding pages talking about the Big Bang, perhaps at least a little time ought to be spent considering what might be entailed by a modern alternative to the Big Bang scenario.

As was the case during the previous eight chapters, the point of this chapter is not to prove or disprove a given cosmological theory. Instead, the intent is to critically reflect on some of the data and arguments that are being used to support a particular position in an attempt to determine what, if anything, such an approach has to offer.

If some alternative cosmological theory turns out to be true, then, as is the case with respect to the Big Bang model, such a truth would not threaten the way I engage the Final Jeopardy challenge that requires me to provide some sort of final response concerning the reality problem. I consider truth, whatever its nature might be, to be an asset rather than a liability.

Before considering the specific cosmological model of Steinhardt and Turok, a little background might be in order. This will include a brief overview of the steady state model that existed up until the mid-1960s and, as well, such background material will take a relatively quick journey into the realm of branes.

-----

Einstein believed the universe operated in a more, or less, stable manner that, notwithstanding local changes, would permit things to continue on in accordance with the principles of physics. He was advocating a sort of quasi-steady state model of the universe.

When he realized that one of the implications of his theory of general relativity involved a fate for the universe that ended in its gravitational collapse, he added the cosmological constant, $\Lambda$, as a means of countering the force of gravitation and, thereby, preserving the long term stability of the universe.

From the perspective of accepted science, there actually was no good reason why the universe had to continue on ... why it couldn't be subject to a gravitational collapse. However, the general conceptual tenor of the early 1900s revolved around the idea of a stable, maybe even a somewhat static – albeit dynamic -- universe, and, therefore, Einstein introduced his fudge factor, Λ, to preserve the assumptions underlying the generally held belief that the universe supposedly was characterized by long-term stability.

During the late 1920s, James Jeans advanced the first version of what, subsequently, would become known as a steady state theory of the universe. Jeans proposed that the universe remained relatively steady through the continuous creation of matter that, somehow, took place.

Hubble's research appeared to indicate that the universe was expanding. In order to counterbalance that expansion, Jeans believed that new material had to be introduced into the universe.

Again, if the universe were truly expanding in accordance with Hubble's findings, then, there is nothing that requires some sort of countervailing property to be added. Just as the fate of the universe might end in a gravitational collapse, so too, the fate of the universe might just go on expanding ... assuming there was no structural limit that, at some point, might constrain or curtail such expansion.

Nonetheless, as previously noted, the idea of long-term stability was present in the conceptual zeitgeist of the times. Einstein had invented Λ to keep the universe stable and steady, while Jeans imagined that there must be some sort of continuous creation of matter taking place that kept pace with the expansion of the universe and, thereby, helped keep the material universe in a, more or less, steady state.

Although the idea of continuous creation of matter became an integral part of the steady state model (and, in a modified form, was even part of a version of the steady state theory that was advanced more than 65 years later in 1993 by Fred Hoyle, Jayant Narlikar, and Geoffrey Burbidge), its presence is not actually necessary. That is, one could conceive of steady state models that do not consider the continuous creation of matter to be an essential feature of the universe.

The amount of material that supposedly had to be continuously created in Jeans' Steady State model turned out to be extremely small, Nonetheless, the idea of matter continuously appearing, as if by magic, bothered the sensibilities of some individuals who believed that matter could neither be created nor destroyed.

However, many of the latter individuals have learned to live with the dynamics of virtual particles that, allegedly, are created and destroyed all the time while, simultaneously, somehow, preserving the laws of conservation. Perhaps, in some similar fashion, matter might be continuously introduced into the universe without -- when considered from an appropriate perspective -- upsetting the laws of conservation.

In order to accommodate Hubble's findings concerning an, apparently, expanding, universe, the Steady State model incorporated the notion of expansion into its theoretical structure. The Big Bang model had not, yet, come into being.

Once the idea of a Big Bang did arise, one of the advantages that a Steady State theory offered its proponents was that they did not have to become bogged down with trying to account for the elusive and unknown dynamics of the Big Bang. The Steady State model eliminated the idea that there was a beginning and an end to the universe, and, therefore, the issue of origins (and its many attendant problems) did not arise in the context of such a theory.

Just as a Steady State model didn't need to contain the notion of a continuous creation of matter in order to be viable, so too, a steady state theory of the universe didn't have to claim that the universe – at least in its present sense -- is endless. The issue of origins is one thing, but the fate of the universe is quite another matter.

Those who are advocates of a Steady State theory of cosmology might be able to find common ground with proponents of Big Bang scenarios when it comes to the future of the universe. There is nothing in physics that requires the universe – whether it began in a Big Bang or always had existed in the past – to go on forever in a manner that is similar to its present status.

The universe as we know it could end in a entropy-induced stupor or in a Big Crunch from which there is no rebound or in the form of an endlessly expanding universe that, eventually, thins matter and energy to such an extent, that nothing happens that is of much interest. Such possibilities might, or might not, occur, but Steady State models cannot rule them out and, consequently, the idea that a Steady State model must propose a universe that goes on forever does not have to be one of that model's structural features.

For much of the 1930s and 1940s, the Big Bang theory remained in the background. It had been invented, drew some interest, and, then, was discarded by most scientists.

Both the Steady State model and the Big Bang theory shared a number of properties. For example, they both featured the idea of expansion, and, therefore, they each interpreted the phenomenon of redshift in the same manner.

Moreover, they both adhered to the cosmological principle. In other words, they each claimed that when viewed through the lens of an appropriate scale, the universe looked pretty much the same no matter what one's frame of reference might be.

They also shared one other feature. At the heart of each theory was a mystery.

For the Big Bang, the mystery was ensconced in the nature of the Big Bang. What existed prior to that event, and what led up to it, and what triggered it?

The mystery inherent in the Steady State model involved the continuous creation of matter issue. There was no physical mechanism to account for that sort of on-going materialization in a way that did not violate various laws of conservation.

Perhaps because of the theological overtones of the Big Bang theory, most scientists in the 1930s and 1940s preferred the Steady State model. The mystery underlying the idea of continuous matter creation apparently was less threatening to them than the mystery surrounding an event that seemed somewhat Biblical in nature.

| Cosmological Frontiers |

An early disagreement between proponents of the Big Bang theory and Steady State advocates appeared to be decided in favor of the latter group of individuals. More specifically, in 1946, George Gamow, working from the perspective of a Big Bang model, sought to explain how various kinds of elements might be created from the intensely hot, dense particle soup that filled the universe shortly after the Big Bang took place.

Gamow, together with his graduate student, Ralph Alpher, and Robert Herman -- who worked in the Applied Physics Lab at John Hopkins -- collaborated to provide a plausible account of nucleosynthesis for the Big Bang theory. The trio did a lot of calculating and arrived at the conclusion that one could explain the relative abundance of elements that were observed in the universe by supposing that protons and neutrons could be captured while speeding about in the hot, dense soup that ensued from the Big Bang and, then, would become fashioned into more and more complex elements if the dynamics of nuclear interaction took place under the right kind of circumstances within the aforementioned hot, dense, particle soup.

Their calculations proved to be fairly accurate with respect to accounting for the relative abundance of hydrogen and helium. However, the theory began to break down when considering the relative abundance of heavier elements such as lithium.

Fred Hoyle, operating through the framework of a Steady State model, proposed an alternative to the account advanced by Gamow, Alpher, and Herman in relation to the origins of chemical elements in the universe. Hoyle believed that such elements were not produced in some hot, dense particle soup that supposedly arose following a Big Bang, but, rather, those elements were synthesized within stars.

In order for more and more complex forms of nucleosynthesis to be realized within the interior of stars, Hoyle believed there had to be a form of carbon (Carbon 12) that had a certain kind resonance (7.69 MeV) that, together with beryllium-8 and helium-4 would permit nuclear reactions (subsequently referred to as a triple-alpha process) to proceed in the direction of heavier elements. In 1952, the sought-for carbon resonance was found and measured and, in doing so, a way had been found that helped to demonstrate how heavier and heavier atoms -- up to, and including, iron -- might be created within stars.

A proponent of the Steady State model – namely, Fred Hoyle – had made a prediction and developed a theory concerning the nucleosynthesis of chemical elements that did a better job of accounting for observed data than advocates of the Big Bang – that is, Gamow, Alpher, and Herman – had been able to do. The scientific advantage belonged – seemingly -- to the Steady State model.

Subsequently, Big Bang theorists modified their model and incorporated Hoyle's ideas into their own theory. Researchers involved with the Big Bang model, then, went on to make certain predictions that were not made by, or consistent with, the Steady State model.

Later on, those predictions appeared to have been empirically confirmed. The theoretical advantage seemed to move in the direction of the Big Bang, and, then after 1965, tended to stay with that model.

The final nail in the coffin of the Steady State model supposedly came in the form of the discovery of Cosmic Microwave Background Radiation in 1965. The Big Bang theory had predicted that such a remnant should be left over from the eponymous event (and even made a number of efforts to quantify it), whereas the Steady State model had made no such prediction.

While the Steady State model might not have predicted the existence of Cosmic Microwave Background Radiation, this did not necessarily mean that such a model couldn't provide an explanation for the presence of that sort of radiation. For example, microwave radiation is likely to be part of any condition of thermodynamic equilibrium that might have been reached over a period of time in a universe that operated in accordance with the properties of a Steady State model.

Considerations of Cosmic Microwave Background Radiation aside, there was an important potential present in the Big Bang theory that did not seem to have a matching counterpart within the Steady State model. The presence or absence of such a potential might play an important role in determining whether the Big Bang theory or the Steady State model better reflected the nature of the universe.

More specifically, from the 1970s onward, almost all, if not all, theoretical research concerning the unification of forces seemed to presuppose the idea that the symmetry breaking event(s) that led from one, overarching force to four or more forces took place in a universe that exhibited extremely high energies and temperatures and that such energies/temperatures apparently played some causal role in triggering various processes of symmetry breaking. A Steady State model, on the other hand, seems to be devoid of such energies/temperatures and, therefore, appears to lack the potential to provide a plausible account for a unified theory of forces.

Of course, if symmetry breaking could occur through conditions that did not involve high energies and temperatures (and no one currently knows what those sorts of conditions might look like), then the absence of such high energies and temperatures (along with concomitant theoretical potentials) is not necessarily a problem. Moreover, if – despite the work that has been done in conjunction with the electro-weak theory of unification – there never was a time when all forces were hidden, as a potential, within the structure of some unified field theory or one, or another, version of Supersymmetry, then, there is no need to provide an explanation for something that never existed (a unified modality from which different forces arose under appropriate conditions) or to provide an account of something that never happened (i.e., symmetry breaking).

Neither the Big Bang theory nor the Steady State Model predicted the existence of dark matter (assuming it actually exists). Furthermore, neither of those two perspectives predicted the existence of dark energy (if it actually exists).

If dark matter and dark energy do exist, neither of those two entities is inextricably tied to either a Big Bang theory or a Steady State model. In other words, one might be able to come up with an account of dark matter and dark energy that made sense within the context of either of those two theoretical frameworks ... especially when one considers the possibility that dark energy – to whatever extent it exists – is not necessarily a function of the Big Bang but could give expression to a phenomenon that is quite independent of the Big Bang.

The Big Bang theory tends to construe redshift in terms of distance, recessional velocity, and/or the stretching of space. However, if intrinsic redshift is a real phenomenon and if, as well, tired light and gravity play a bigger role in redshift than previously believed, then, the redshift lens through which the Big Band filters many cosmological phenomena might be giving a distorted picture of what is taking place in the universe.

Yet, even if redshift is largely a function of: Distance, recessional velocity, and/or the stretching of space, there is nothing present in such an understanding that couldn't be reconciled with an appropriately amended version of a Steady State model of the universe. In short, the latter perspective is capable of accommodating the idea of redshift as well as the Big Bang theory.

Most mainstream astrophysicists and astronomers prefer the Big Bang theory to the Steady State model. Such scientists seem to believe that the Big Bang theory provides a much, much better account of cosmology than does the Steady State model.

Nonetheless, the foregoing preference might be more a function of psychological, sociological, political, and educational forces than it is a reflection of whether either of those two models constitutes a more accurate reflection of the nature of reality. Indeed, scientifically speaking, such a preference is somewhat puzzling given: (a) The many parallels and areas of overlap that exist in relation to those two accounts; (b) the phenomena that neither theory predicted (e.g., universal constants, the asymmetry between matter and anti-matter, dark matter, dark energy, black holes); (c) the ability of both frameworks, if required, to work with issues involving redshift, expansion, nucleosynthesis, cosmic microwave background radiation, dark matter, dark energy, the lifecycle of stars, and so on; (d) the numerous unresolved cosmological problems that continue to persist in conjunction with both theoretical frameworks.

-----

Before moving on to consider the cosmological perspective of Paul Steinhardt and Neil Turok, the idea of branes and associated notions need to be explored. Branes play an important role in the theory being advanced by the two aforementioned researchers.

Perhaps a good way to begin our journey toward the theoretical world of branes is to begin with a brief discussion of gravity. More specifically, Newton's inverse square law (i.e., gravity is proportional to the square of the distance between two objects) is a reflection of the three spatial dimensions through which it is expressed or manifested.

For instance, if the universe involved just two dimensions, the force of gravity in, say, a circle would have been distributed across that circle, and, as a result, the strength of the gravitational force would have decreased at a rate that is slower than what occurs in the context of three spatial dimensions.

On the other hand, if the universe gave expression to more than three spatial dimensions, then gravity would be spread across the surface of a hypersphere (n-dimensional counterpart to a three dimensional sphere). As the surface area separating objects immersed in a hypersphere became larger, the strength of gravity exhibited in such circumstances would fall off much more quickly than would occur in a universe consisting of just three dimensions.

If, as various versions of string theory and Supersymmetry claim, we live in universe that involves more than three spatial dimensions, then, how does gravity work its way through those extra dimensions so that we observe the inverse square law in the everyday world of three spatial dimensions? According to such theories, the spatial dimensions that exist beyond the three with which we are familiar are compactified, and, therefore, quite small.

As gravity spreads through those compactified spaces it is bent by the boundaries of the dimensions through which it is spreading. Moreover, the compactified spatial dimensions serve to channel or direct that gravity toward the three visible spatial dimensions.

While the force of gravity is calculated to spread out radially over the tiny distances involved in compactified spatial dimensions, nonetheless, in relation to the longer distances entailed by the three visible spatial dimensions, what happens on the level of compactified spatial dimensions becomes – mathematically speaking – relatively negligible as far as the dynamics of three-dimensional space are concerned. Consequently, the properties of three-dimensional space assert themselves, and the inverse square law of Newtonian gravity with which we are familiar dominates what transpires within such a multi-dimensional context.

The impact of compactified spatial dimensions on the strength of the gravitational force will only be tangibly manifest in relation to distances that are smaller than the size of the compactified dimensions through which the gravitational lines of force are spreading. When the spatial distance separating two objects is larger than the size of those compactified dimensions, Newton's inverse square law holds.

Many theorists believe that the size of the compactified, curled up spaces is on the order of a Planck-length. This involves a scale that is $1.616 \times 10^{-35}$ meters in size.

Such a scale is somewhat arbitrary. Compactified spaces might be larger or smaller than the Planck length … no one really knows because no one has been able to peer down into the realm where compactified dimensions might exist.

Some theorists have raised questions about how, or if, compactified dimensions interact with three-dimensional space … assuming, of course, that the latter space is made from an assemblage of digital units of some kind. If compactified dimensions were able to interact with the quantum units of three-dimensional space, then questions would arise concerning the nature of the dynamics that might occur along partially shared boundaries involving various compactified dimensions and three-dimensional space.

Of course, one possibility that does not seem to have been taken into consideration by string theorists and proponents of Supersymmetry is that the extra dimensions to which allusions are being made might not be spatial in nature. Qualitatively speaking, dimensions might be much more akin to time than to space … that is, time seems to be something other than spatial in character and, therefore, so too, extra dimensions might be non-spatial in nature as well.

Representing time by means of a spatial co-ordinate value tends to "spatialize" the dimension of time … that is, the process of spatializing filters time through, and frames time in terms of, the biases of space. Spatializing time might permit various kinds of useful quantitative calculations to be made, but this is done at the cost of distorting the way in which time and space give expression to different kinds of dimensional properties.

If extra-dimensionality is not construed in spatial terms, then, the impact, if any, that such extra dimensions have upon the strength of gravity (or any other force) will be a function of the extent to which -- as well as in what way, if at all -- the qualitative properties of those dimensions are sensitive to the presence of the gravitational force (or other forces). Dimensions that are not sensitive to the presence of gravity (or other forces) will have no effect on the inverse square law of Newton (or any other kind of physical law) that tends to be operative within the context of three spatial dimensions.

Moreover, extra dimensions that do not involve spatial properties do not become entangled in the problems confronting extra dimensions that are spatial in nature. The notion of compactified space arose as a means of accounting for why we can't see the extra dimensions of space, but if whatever extra dimensions that are being proposed were not spatial in nature, then, there is no need to try to explain them away through arbitrary notions such as compactified spaces.

Although we can observe the effects of time, we don't see time directly. Nevertheless, no one feels compelled to explain why time is not visible in the same way that theorists apparently feel obliged to account for why extra spatial dimensions might not be visible.

Notwithstanding the foregoing considerations, a brane is the generalization of a point-particle to higher spatial dimensions. If one starts with a point particle, then a point gives expression to a brane of dimension zero.

The term "brane" is derived from the word "membrane". A membrane is a brane of dimension two, while the basic unit of string theory – i.e., a one-dimensional, vibration-laden object – is a brane of dimension one.

Various kinds of characteristics can be built into the notion of a brane. For instance, branes can be provided with features such as mass, charge, and other quantum properties.

In addition, branes can be given the capacity to trap forces and particles on their surfaces. Furthermore, branes can be supplied with the sort of qualitative and quantitative attributes that are needed to operate in accordance with the principles of quantum mechanics.

From the perspective of brane theory, the three-dimensional world in which we live can be conceived of as a three-dimensional slice of a higher dimensional reality and is referred to as a 3-brane. In general, any given kind of brane represents a dimensional slice of some kind ... either within, or bounding in relation to, some given higher dimensional object.

Dimensions don't have to be curled up or compactified. They can be finite constructions that terminate on a boundary of some kind that represents the furthest reaches of that kind of dimensionality.

For example, given a certain kind of higher-dimensionality, branes can be considered to be the boundaries that mark the extent to which such higher dimensionality participates in the complete realm of spatial dimensionality that is known as the "bulk". The aforementioned kind of higher dimensionality is but a slice of the bulk.

The way in which the bulk manifests itself through a given context of higher dimensionality (which is a lower dimensional object relative to the bulk) is marked by branes. Branes give expression to boundaries through which a given modality of higher-dimensionality is linked to the still higher dimensional 'bulk' ... that is, the entire realm of possibilities involving spatial dimensionality.

The bulk extends in all possible spatial directions. The brane only extends in accordance with some of those spatial possibilities as a function of the properties that are inherent in the kind of brane being considered.

Among other possibilities, branes exhibit what is known as "reflexive boundary conditions". In other words, if a force or particle encounters a brane, then, that physical entity bounces off the brane without energy being lost, absorbed, or leeching through the boundary to which the brane gives expression.

Not all branes mark boundaries. While a brane that constitutes a boundary is considered to have a lower dimensionality than the dimensional object it bounds, a brane that does not form a boundary has higher dimensionality spaces on all sides of it, and, therefore, still constitutes a lower dimensional object relative to the context in which it is situated.

Branes have the capacity to restrict movement as a function of their spatial properties. In fact, this restrictive dimension of branes is what differentiates them from a theory that just involves the idea of multidimensionality.

The dimensionality of a brane is rooted in the foregoing property of restricted movement. More specifically, the dimensionality of a brane is determined by the number of degrees of freedom that an object (particle, string, or force) has to move in conjunction with that brane.

A 2-brane has two degrees of freedom with respect to movement. A 3-brane has three degrees of freedom, and so on.

Some objects (particles, strings, or forces) are vulnerable to the ways in which various kinds of brane can restrict movement while other objects are not so sensitized. The latter objects are free to move throughout the full realm of the bulk (i.e., all possible spatial dimensions).

Physically speaking, the inhabitants of Edwin Abbot's 1884 literary creation of *Flatland: A Romance of Many Dimensions* could only move in two directions, while the three-dimensional interloper who also was featured in that tale could move in three directions. On the other hand, given that the beings of Flatland apparently were self-aware as well as were able to feel, think, imagine, and speak, then, apparently, the latter sorts of qualities were not necessarily restricted with respect to the possible directions in which they might conceptually move ... although the physical restrictions in movement did seem to have an impact on the kinds of theoretical issues some of them were prepared to consider.

Particles, strings and forces that have been trapped or restricted by the properties of a given modality of brane, can only influence other particles, strings, and forces that are similarly trapped or restricted by that same brane. Such particles, strings, and forces are brane-bound.

However, there are some forces, such as gravity (the starting point for this section on branes), that are brane-independent. According to the theory of general relativity, gravity is inherent in the very fabric of spacetime, and, therefore, its lines of force have the capacity to spread throughout every spatial dimension.

Neither Einstein nor anyone else has demonstrated how gravity is woven into the fabric of either spacetime or space and time. The claim that gravity is woven into the fabric of spacetime or time and space is, at the present time, little more than an assumption.

In other words, the general theory of relativity constitutes a behavioral description of the way in which gravity manifests itself in a given physical context. That theory says nothing about the specific, concrete nature of the phenomenon (i.e., gravity) that makes such behavioral properties possible.

While it might be the case that gravity has the capacity to spread through every spatial dimension – compactified or otherwise -- there is nothing (other than the aforementioned assumption) that necessarily requires gravity to penetrate all dimensions – spatial or otherwise. Perhaps, some dimensions are not vulnerable or sensitive to the presence of gravity, and, if so, then, gravity either will not be able to penetrate those spatial dimensions or will do so in a way that is devoid of interactional dynamics.

According to many proponents of brane theory, gravity is the common currency of all spatial dimensions. Gravity is what links the bulk (the full extent of possible spatial dimensions) and all lesser dimensional spaces.

In addition, there might be other forces and particles flowing through the bulk that are capable of interacting with an array of branes just as gravity does. If this is the case, then branes and the bulk have the potential to be connected by means of such forces and particles as well.

The idea of a braneworld involves a brane of n-dimensions that places restrictions on the degrees of freedom available to particles and forces that are trapped by such a brane. Such branes are the lower dimensional boundaries or surfaces for higher dimensional spaces.

For example, from the perspective of the theory of branes, human beings live on a 3-brane universe. Most of the "objects" of that brane (such as planets, stars, galaxies, life forms, and so on) are restricted by the degrees of freedom that govern the dynamics characteristic of that

3-brane world, while there are some objects, such as the force of gravity, which spread through that brane and the bulk without restriction.

Currently, theorists have no idea how many possible kinds of branes exist in the bulk. Moreover, they have no idea about the extent of the dimensional possibilities that make up the bulk.

The term "multiverse" is often used in relation to a cosmos or universe that contains more than one modality of brane. Furthermore, generally speaking, the branes that exist within such a multiverse are considered to interact with one another either very weakly or not at all.

Each of the branes within a multiverse operates in accordance with its own set of properties and dynamics. The geometric possibilities that might exist in conjunction with such branes, as well as the way in which those possibilities fit into the overall geometric possibilities inherent in higher dimensional spaces, are indefinitely great.

In 1989, Joseph Polchinski, along with Rob Leigh and Jin Dai, were exploring the structural properties of string theory (Petr Hořava, a Czech string theorist, was working independently along the same lines), and they uncovered the presence of a particular kind of brane known as a D-brane. The 'D' is in honor of Peter Dirichlet, a German mathematician from the 1800s whose work on boundary conditions helped orient the investigations of Polchinski, Leigh, Dai and Hořava.

Open strings have free ends. The aforementioned four investigators discovered that the only permissible place where such ends might terminate is on a brane that gives expression to a particular set of boundary conditions – namely, a D-brane.

The ends of a string don't have to terminate on the same brane. However, whatever brane the open end of a string links up with, that brane will be a D-brane.

As noted earlier, branes can be given a variety of properties. These include: Dimensionality, shape, charge, mass, size, movement, modes of interacting, and tension.

One kind of brane, known as a p-brane, gives expression to certain kinds of solutions to the equations of general relativity. These solutions have been mathematically explored by, among others, Andy Strominger.

Strominger discovered that when p-branes wrap around tiny, compactified areas of space, those branes exhibit properties that resemble a massless particle. In other words, within the context of certain kinds of spacetime geometries, new particles are made possible through the presence of the right kind of branes, and, therefore, Strominger opened up the possibility that not all particles are necessarily a function of vibrations in strings.

In 1995, Polchinski demonstrated mathematically that branes are integral to string theory. Among other things, he showed that p-branes and D-branes are transformable into one another at energies for which the theory of general relativity and string theory make similar predictions.

Unfortunately, neither Polchinski nor anyone else has been able to show -- in a determinate manner -- how string theory would delimit the properties of branes (and vice versa) in a way that would give rise to the sort of three-dimensional phenomena and experiences with which we are all familiar.

A possible preliminary breakthrough involving branes and strings eventually bubbled to the surface of awareness in 1995. Ed Whitten showed that in conditions involving low energies, there existed a kind of ten-dimensional superstring theory exhibiting a property of strong coupling that was equivalent to eleven-dimensional supergravity.

As the name suggests, superstring theory involves strings. However, eleven-dimensional supergravity contains no strings, but, instead, employs 2-branes.

Nonetheless, if one of the eleven dimensions of supergravity is curled up and wrapped up in a 2-brane, the combination has a string-like appearance to it. That is, like a string, the 2-brane that is wrapped around a curled up dimension seems to exhibit only one dimension.

Thus, under certain conditions (i.e., low energies, strong coupling), eleven-dimensional supergravity theory is equivalent to ten-dimensional superstring theory. Despite their superficial differences, underneath it all, ten-dimension superstring theory and eleven-dimension supergravity display the property of duality in which two seemingly different frameworks give expression to the same sort of theory.

Throughout 1995 and 1996, a variety of theorists contributed to enhancing the idea that was inherent in the insight that a number of ten-dimensional frameworks were dual with the eleven-dimensional theory of supergravity. Prior to such research, many investigators believed there were five different editions of string theory, but, as it turned out, there was just one duality in which five, ten-dimensional string theories were shown to be all equivalent with one another as well as with a eleven-dimensional supergravity theory that, at first glance, appeared to contain no strings.

The general idea for Ed Whitten's so-called M-theory arose out of the foregoing perspective. He believed there must be some overarching mathematical framework that encompassed all of the ten-dimensional editions of string theory, together with eleven-dimensional supergravity theory, quite apart from condition of low energy.

M-theory might be the means through which a more complete theory of superstrings could be developed. In addition, if fully realized, M-theory might also lead to a viable theory of quantum gravity.

According to Whitten, every facet of eleven-dimensional supergravity has a matching counterpart in ten-dimensional superstring theories. More specifically, the momentum of any given object in eleven-dimensional supergravity requires eleven numbers to describe its location in spacetime, and, therefore, if there is to be a counterpart in ten-dimensional superstring theory, there must be an equivalent set of numbers to specify the value of the superstring counterpart to supergravity.

The extra number that is needed in ten-dimensional superstring theory involves charge. Particles (strings) that carry a charge in ten-dimensional superstring theory are known as Do-branes, and Do-branes correspond to the property of momentum in eleven-dimensional supergravity.

Particles exhibiting eleven-dimensional momentum in supergravity behave like charged Do-branes in ten-dimensional superstring theory. Thus, despite their apparent differences, both ten-

and eleven-dimensional theories end up giving expression to comparable behavioral properties.

The notion of branes helped pave the way to realizing that dualities exist between ten-dimensional string theories and eleven-dimensional supergravity. Nonetheless, branes carry a cost.

Branes have the capacity to trap forces and particles in ways that were not anticipated by string theory when the latter idea first surfaced. Furthermore, branes can assume many orientations within the context of higher dimensions.

Strings and branes have a synergistic capacity to interact in many different ways in a variety of contexts with respect to contexts of higher-dimensionality. Consequently, the combinatorics inherent in such dynamic interfacing tends to become overwhelmingly large.

Branes have been shown to be integral to the nature of string theory. However, no one knows whether, or not, either branes or strings are integral to deriving an accurate description of various aspects of reality.

Branes and strings are mathematical constructs. Their value as a potential means through which to develop viable frameworks for describing observed phenomena is only as good as the capacity of such constructs to generate models that accurately reflect this or that facet of physical reality.

Braneworlds are an exercise in modeling that brings strings and branes together in, hopefully, heuristically valuable ways. Ed Whitten and Petr Hořava produced the first braneworld.

Their braneworld involves eleven dimensions. Those dimensions bound two parallel branes, each of which consists of nine spatial dimensions that bound a bulk involving eleven dimensions of spacetime.

The foregoing branes have been fitted with an array of forces and particles that can interact with one another in ways that are shaped and constrained by the nature of the branes themselves. Moreover, in the Hořava-Whitten braneworld, five of its dimensions are curled up or compactified in some fashion (i.e., give expression to Calabi-Yau manifolds).

The properties of the Hořava-Whitten braneworld are sufficiently rich, nuanced and complex to be able to provide ways of describing the forces and properties that inhabit the Standard Model of quantum mechanics. Their braneworld is also able to accommodate a force of gravity.

One of the problems with the Hořava-Whitten braneworld – and this problem carries over to many, if not all, of the braneworlds that have arisen since the original Hořava-Whitten braneworld thought experiment was conducted -- is that experimentally testing such models is difficult, if not (at least for the foreseeable future) impossible, to do. Many of the dimensions of the Hořava-Whitten braneworld are very tiny and compactified, and, consequently, trying to empirically determine the size, shape, and properties of those spaces (if they actually exist) is beyond our present technological reach.

Constructing braneworlds is an exercise in playing around with branes and strings in order to see what might happen. In effect, branes, with whatever set of properties one likes, are available for hire and can be used to construct models that – if one is very, very, very lucky – possibly (even if improbably) might give rise to a way of describing things that is applicable to reality.

However, assuming one were sufficiently fortunate to be able to happen upon a heuristically valuable system that was capable of arranging branes and strings into a powerful system of description, nevertheless, being able to accomplish this does not mean that one will necessarily understand the nature of what is being described. Nor does such a system of description necessarily mean that one will be able to understand why things operate in the way they do rather than in some other fashion.

After all, both Newton and Einstein devised systems that were capable of describing the effects of gravity, but neither of them understood the nature of that (i.e., gravity) which made such effects possible. Similarly, quantum physicists have long acknowledged that while they have developed a very powerful system for describing, in a very precise manner, the behavior of quantum events, no one really understands what is taking place.

In short, as was true in the case of Searle's Chinese Room conundrum, the ability to use description in correct ways does not necessarily entail an understanding of what is being described. Moreover, reality might consist of far more than what our methodologies enable us to describe.

-----

Some proponents of the Big Bang scenario claim – and manage to do so with a straight face – that the Big Bang created time and space. How a physical event manages to give rise to the very conditions that it seems to presuppose is something of a mystery.

Guth once referred to the alleged benefits that ensued from inflationary theory as constituting the ultimate free lunch. Using the Big Bang to account for the existence of time and space would appear to take care of providing a free breakfast, brunch, supper, and a late-night snack.

At first glance, Paul Steinhardt and Neil Turok seem to feel a certain willingness to pay for their theoretical meals. For example, they don't necessarily believe that the Big Bang brought time and space into existence (although they seem to be open to the possibility that this might have happened), and, moreover, they maintain there was more than one Big Bang … perhaps an endless number of them.

Steinhardt and Turok are proponents of a cosmological theory that is cyclical in character. In other words, they maintain that the universe goes through a series of lifecycles in which the end of one lifecycle helps bring about the beginning of the next lifecycle.

Steinhardt and Turok argue that the foregoing cyclical model produces results that are capable of reflecting all of the astronomical data (such as WMAP) that has been collected over the last century, or so. One might also point out that the cyclical model appears to multiply the number of problems that plague the original Big Bang scenario by a factor that is equivalent to the number of cycles that have taken place over the many alleged lives of the universe.

One of the features that the foregoing cyclical theory does not share with the Big Bang scenario is the latter's contention concerning an early, brief, but intense period of inflation in which the universe expanded by a factor of $10^{100}$ in $10^{-30}$ seconds. Although, initially, Paul

Steinhardt had supported the idea of an inflationary universe, he has since come to the conclusion that the notion of inflation is far too artificial and ad hoc since it has been modified again and again as new possibilities show up (e.g., dark matter, dark energy), and, as well, the notion of inflation involves an expansive energy that somehow turns on and off in inexplicable ways.

Both Steinhardt and Turok trace the primary impetus for the development of their cosmological theory to a lecture that was being given by Burt Ovrut. During the talk, Ovrut was talking about an idea that had been introduced by Ed Whitten and Petra Hořava in relation to string theory.

The idea described the manner in which two different realms could be separated by an extra dimension. Each realm could have its own system of dynamics, as well as properties of matter, radiation and force.

The distance separating the two realms might be extremely miniscule. However, the rules of string theory prevented those two realms from interacting with one another across the extra dimension that separated them.

After the talk, Steinhardt and Turok converged on the speaker from different directions. They began peppering Ovrut with questions as well as exploring possibilities with him.

One of the main themes running through their questions and speculations had to do with whether or not a collision between the two realms mentioned by Ovrut might constitute a Big Bang event. In addition, they explored the possibility that if there had been no early burst of inflation, then, perhaps, evidence might still be in existence (i.e., it was not swept away by inflation) concerning the properties of such a collision.

The Big Bang is expressed in terms of the Standard Model of quantum mechanics. The cyclical theory of Steinhardt and Turok is rooted in string theory.

Whatever problems the latter model shares with the Big Bang scenario, it also harbors an additional difficulty. More specifically, it presupposes that string theory is something more than an amalgamation of interesting mathematics and that string theory has something to do with the real world. Yet, at the present time, string theory is beset with difficulties when it comes to formulating a coherent hypothesis concerning the real world and doing so in a way that can be tested empirically.

For the moment, let's leave aside whatever problems plague string theory. Let's consider some of the data with which both the Big Bang theory and the cyclical theory of cosmology work.

Astronomers and astrophysicists use different instrument packages to generate data that is used to help construct a picture of the universe. For example, the Sloan Digital Sky Survey uses images that are photographed by means of a camera that contains a charge-coupled device (CCD) that is very sensitive to the presence of light coming through a 2.5meter telescope in New Mexico.

The images generated through the foregoing process have been assembled into a collage that covers (in slices) a quarter of the Northern Hemisphere sky (A similar survey has been conducted in conjunction with the Southern Hemisphere). More than 2 million galaxies are contained in the Northern Hemisphere collage and the depth of that data field extends out to a billion light years.

Despite its early imaging problems, the Hubble Telescope has also played an important role in helping to map the cosmic heavens. For instance, by programming the Hubble telescope to focus for ten days in a particular direction of the universe (which, initially, appeared to be devoid of any stellar or galactic inhabitants) the Hubble Deep Field image was created, and this depicts thousands of galaxies with varying sizes and shapes that existed billions of years ago.

The data that was gathered by WMAP (Wilkinson Microwave Anisotropy Probe) supposedly comes from the Cosmic Microwave Background Radiation that was released some 380,000 years after the Big Bang when the universe cooled down sufficiently to permit electrons and charged nuclei to decouple from the opaque, radiation soup that had dominated the universe from the time of the Big Bang ... a decoupling that made the universe transparent to the passage of light and that, subsequently, might be captured through the sort of instrument packages that underlie the Sloan Digital Sky Survey or the Hubble Deep Field image.

The Cosmic Microwave Background Radiation gives expression to a phenomenon that started to pump out its radio noise nearly 13.5 billion years ago. Because of the opaque properties of the radiation soup that dominated the universe up until about 380,000 years following the alleged origins of the universe, proponents of the Big Bang believe there might not be any way to gather data with respect to the universe prior to the time when Cosmic Microwave Background Radiation began to make its presence known.

A large part of our "understanding" – if that is what one can call it -- concerning cosmology comes by way of inference. For example, one takes data from WMAP, the Hubble Deep Field image, the Sloan Digital Sky Survey, along with data from many other sources and, then, individuals make inferences about how they believe that data came to have the properties it is observed to have.

For example, the data of WMAP constitutes an empirical basis for making an inference there was a Big Bang that, eventually, led to the emitting of Cosmic Microwave Background Radiation and, therefore, serves as a marker for that prior event. However, if the Big Bang never took place, and if Cosmic Microwave Background Radiation gives expression, instead, to the presence of the sort of thermodynamic equilibrium that might arise through the on-going dynamics that are – and have been -- transpiring in the universe as a whole for some time, then, one cannot use the data of WMAP to infer the prior occurrence of a Big Bang ... that data is actually evidence for some other kind of cosmological phenomenon.

Similarly, the data from the Hubble Deep Field image and the Sloan Digital Sky Survey (along with other sources of astronomical data) is used as a basis for generating inferences concerning how stars and galaxies initially formed, as well as a basis for making inferences concerning the nature of the lifecycle of stars and galaxies. However, rather that demonstrating that more and more primitive forms of stars and galaxies dominate the universe the further back one probes into the past, the available evidence indicates that even the most distant galaxies are comprised of stars that exist in different stages of their lifecycle, and, therefore, the Standard Cosmological Model doesn't seem to be able to explain why that sort of complexity is present at such an, allegedly, early point in cosmological development.

According to the most recent measure for the age of the universe (European Space Agency Planck mission), the universe is said to be about 13.82 billion years old (This is actually a median value, and according to data from the Planck mission, the universe might be slightly older or slightly younger by some 120 million years). The foregoing age includes the period of some 380,000 years during which the universe supposedly was opaque to the transmission of light throughout the universe due to the way light was scattered by the presence of a dense, hot soup dominated by high-energy radiation.

The 13.82 billion years figure is an inference based on an array of data. That data doesn't necessarily demonstrate that a Big Bang occurred 13.82 billion years ago but, instead, the latter figure could indicate that the data we have available to us only takes us back 13.82 billion years and that we don't necessarily know what happened, if anything, prior to that point in cosmic history ... that is, we can only make inferences about what might have transpired prior to the time that falls beyond the horizons of our present temporal measurements.

What happened 13.82 billion years ago? Both Big Bang advocates, as well as Steinhardt and Turok, believe that some sort of cosmic explosion took place that has been shaping our universe ever since.

As was pointed out in several previous chapters, many unresolved problems permeate the perspective of cosmologists in relation to the nature of the singularity that supposedly existed prior to the Big Bang. Additional unresolved questions arise in conjunction with determining both the nature of the Big Bang as well as the nature of the inflationary period that allegedly occurred early on during the period of the Big Bang.

The unknown nature of dark matter and dark energy (assuming that they actually exist) also muddies the waters of Big Bang cosmology. Moreover, problems that are entailed by the redshift issue obscure the situation still further because redshift is crucial to the measurement and interpretation of so many features of Big Bang cosmology.

Notwithstanding the foregoing considerations, Steinhardt and Turok hypothesize that each cycle of the universe consists of six stages. Although each of these stages outlines a picture that is different in certain respects from the one generated through Standard Big Bang cosmology, nevertheless, there are still problems that are inherent in each of the proposed Steinhardt/Turok stages.

(1) A Big Bang occurs in which densities and temperatures are high -- but they are not infinite as they are in the Standard Model of Cosmology – and, therefore, the Steinhardt/Turok Big Bang is governed by physical laws involving finite energy densities in which time continues to flow (i.e., it is characterized by a before and after). Whether the collision of branes could produce the sort of high densities and temperatures required by Steinhardt and Turok to make their theory work is an open question ... one that will be addressed again a little later on.

(2) While there is no period of inflation present in the Steinhardt/Turok cyclical cosmological model as there is in the Standard Model, the Big Bang that is envisioned by Steinhardt and Turok does lead to a set of conditions in which the universe is dominated by radiation just as in the Standard Model of Big Bang cosmology. According to the Steinhardt/Turok model, asymmetries between matter and antimatter occur during this stage, but no account is given for how such an asymmetry arises ... other than to say that the asymmetry is the end result of a process of mutual annihilation that, after the dust settles, gives expression to a universe that consists largely of matter.

(3) The Steinhardt/Turok model then enters into a stage that lasts approximately five million years. During this period of time, stars and galaxies are formed through the influence of gravity, but one has difficulty reconciling the complex structures that are present in the universe (in the form of superclusters and voids that have dimensions involving hundreds of millions of light years) that would have taken far longer to form through just gravitational attraction that, supposedly, has been transpiring for nearly 14 billion years.

(4) Next is a period that is marked by the advent of the influence of dark energy. During this period, the universe is reduced largely to a uniform, diffuse, sea of energies and particles without appreciable, if any, structure, but, as was discussed in 'Chapter 8: Expanding Horizons', there are many good reasons to question whether, or not, dark energy actually exists.

(5) During this stage -- which lasts for a fairly arbitrary period of one trillion years -- dark energy undergoes a process of decay that, supposedly, is both gentle and smooth. The decay of dark energy causes the rate of expansion to slow down and, eventually, this leads to a period of gravitational contraction.

Even assuming that dark energy exists there is no evidence to indicate that dark energy should be subject to decay. Moreover, there is no evidence to indicate that such decay, if it did occur, would be smooth, gentle, or require a trillion years to take place ... this is all just a form of speculation that serves the needs of the cyclical aspect of the Steinhardt/Turok model.

(6) The period of contraction that started in stage (5) transitions into a Big Crunch. During this process, some portion of the dark energy supposedly transforms into energetic radiation and hot matter that fuels a Big Bang and subsequent expansion.

How a form of energy that previously only affected space suddenly becomes able to generate particles and radiation is unknown. Furthermore, although Steinhardt and Turok claim that the state of the universe is smooth and flat both before and after such a Big Bang, this claim is more of an assumption than anything else since there are many factors that might determine whether, or not, such a Big Bang and ensuing expansion were smooth and flat rather than rough and not flat.

Steinhardt and Turok also claim their model provides a way to account for the inhomogeneities that will seed the large-scale structures that will form during the next lifecycle of the universe. Such inhomogeneities supposedly are a function of quantum fluctuations, but as was pointed out in the last chapter of the present book as well in various places during the discussions encompassed within *Final Jeopardy: Physics and the Reality Problem, Volume II*, the notion of quantum fluctuations might be more of a conceptual and mathematical bag of tricks than a reflection of the nature of reality.

Quantum fluctuations are a way of trying to give the impression that one knows something about the character of ontology when this is not necessarily the case. Instead, one is merely sweeping one's ignorance beneath a carpet that alludes to many possibilities in a very oblique and speculative manner.

Steinhardt and Turok maintain that every Big Bang will lead to different results because the underlying quantum fluctuations are random in nature and, therefore, are governed by the laws of chance. This is just another way of saying they have no idea why things turn out the way they do.

There are no laws of chance that govern randomness. Randomness and laws of chance are antithetical to one another.

Throwing dice, dealing cards, and flipping a coin operate in accordance with the laws of chance because those activities are marked by determinate parameters of possibility. Random phenomena contain no such markers.

If there are laws governing the shape of a phenomenon, then, randomness cannot be involved. The very essence of a random phenomenon is that there is no algorithm capable of describing such a sequence and, consequently, there are no law-like properties that are present in that phenomenon.

Steinhardt and Turok are correct to question the peculiarities of inflationary aspect of the Standard Model of cosmology. The current theory of inflation has been cobbled together from a largely disparate array of elements involving: The Big Bang, inflation, dark matter, and dark energy ... none of which – at least as far as is currently understood – necessarily entail one another or empirically imply one another, but all of which rest on theoretical foundations that are constructed from very questionable assumptions.

Unfortunately, Steinhardt and Turok don't appear to have subjected their own model to a similarly rigorous process of critical reflection. As has been briefly outlined during the previous three, or so, pages, each stage of their six-part cyclical model contains significant problems.

When I was 7 or 8 years old, my friend from next door and I would go to the movies on Saturday afternoon, and prior to the start of the show, we would entertain ourselves by asking one another which of two possible gruesome deaths would one prefer to be the means of one's demise. Being asked to choose between the Standard Model of cosmology and The Steinhardt/Turok model seems like being confronted with the task of having to choose between whether one would prefer to sink on a ship that was full of logical and empirical holes of one kind rather than another.

One can play around with energy density curves and see what happens when various parameters (involving, for example, a Higgs field, quantum fluctuations, dark energy, strings, Calabi-Yau spaces, branes, and so on) are altered in various ways. When playing about in this fashion, an individual might even happen upon modes of descriptions that are capable of mirroring some of the empirical observations that have been made in astronomy.

However, given such success, one is faced with the fine-tuning problem. In other words, one must come up with an explanation for how ontology, on its own, was able to bring about a set of conditions that have been fine-tuned in the way that theorists have done somewhat arbitrarily in relation to the parameters with which they have been playing, and, to date, no one has been able to accomplish this.

As noted earlier, a primary impetus for the cyclical cosmological model developed by Steinhardt and Turok was a talk given by Burt Ovrut, a string theorist. One of the theoretical possibilities that arose in conjunction with that talk was the idea that colliding branes might give rise to a Big Bang.

A brane, as previously indicated, is the generalization of a point-particle to higher spatial dimensions. Branes are integral to string theory, and branes have helped lead to the realization that dualities exist between ten-dimensional string theories and eleven-dimensional supergravity.

Branes can be outfitted with all manner of properties involving: Mass, charge, dynamics, dimensionality, quantum properties, size, or shape. In addition branes can be involved in all manner of relationships with higher and lower dimensional objects.

According to Steinhardt and Turok, if two branes collided, then, presumably, this would involve the generation of some kind of dense field of energy. This in turn, supposedly, would lead to the distribution of hot radiation within both branes.

What happens when two branes collide (assuming that branes constitute more than a mathematical construction) depends on many considerations. For example, if the surfaces along each brane exhibit reflexive boundary conditions, then, whatever energy, particles and properties are present in each brane will not be lost, absorbed or leaked, but there is no guarantee that such a collision will lead to a Big Bang.

Moreover, the amount of hot radiation that might be generated would depend on the force with which the two branes collided. In addition, the amount of hot radiation that was produced would be a function of the properties that had been built into the branes being considered.

With the right sort of collision velocities and the right set of properties built into the branes, then a lot of things are possible. However, none of this means that such collision velocities ever actually occurred or that the right kinds of properties were ontologically present within the respective branes (assuming that the latter constitute an accurate description of anything beyond the horizons of mathematics).

Another factor that might affect the outcome of two branes colliding would involve the dimensionality of each brane. Dimensionality, compactification, and the shape of Calabi-Yau spaces can all affect how the energy that might be generated through such a collision could shape the nature of what is distributed and how things are distributed.

Steinhardt and Turok claim that all but one of the six stages of their cyclical theory can be described through generally accepted techniques and modes of calculation. The stage that is not amenable to such techniques and modes of calculation is the first one involving a Big Bang.

Branes exhibit the property of flexibility. Supposedly, when they approach one another, their boundaries each undergo quantum fluctuations that cause the branes to ripple like sails in a shifting wind.

Quantum fluctuations that, supposedly, are governed by the laws of chance are an assumption. If one likes, two branes that are approaching one another can each be considered to be rippling in accordance with quantum fluctuations, but this is an assumption as well.

According to some theorists, heterotic M theory (involving closed strings or loops that are hybrids consisting of superstrings and bosonic strings) contains spring-like forces that arise between branes. Such spring-like forces have been hypothesized to give expression to the interaction or movement involving branes that might, in turn, give rise to a Big Bang.

Whether such spring-like forces actually exist in the world beyond mathematics is a separate issue. Furthermore, even if those forces exist, whether, or not, they are sufficiently powerful to generate a Big Bang is unknown.

Like Icarus, models fly on the wings of their assumptions. Yet, if such wings fly too close to the Sun's reality, those wings might melt away to nothing, thereby endangering the model that such wings, heretofore, have been enabling to soar.

M theory involves complex equations that deal with dimensionality, compactification, branes, and some version of supergravity or quantum gravity. Moreover, M theory is not even fully delineated, so, whatever solutions are generated through M theory tend to be, at best, partial in nature.

To claim that heterotic M theory allows for spring-like forces that might generate the sort of energies that could subsidize a Big Bang is not necessarily saying a lot. Indeed, as an old English proverb indicates, 'There's many a slip twixt the cup and the lip' ... theoretically speaking.

Steinhardt and Turok envision two branes consisting of three, visible dimensions that extend out to infinity. The branes are parallel to one another, and, initially, a considerable distance separates the two branes from one another.

The meaning of "considerable distance" in the foregoing paragraph might be a fairly relative thing. Braneworlds often are described in terms of being separated from each other by distances as small as between $10^{-28}$ and $10^{-30}$ centimeters.

The foregoing distance of separation between branes is set by matching the strengths of gravity and other forces within particle physics in a given context of extra spatial dimensions and compactified spaces in relation to various properties that have been built into such brane pairs. However, if the universe does not consist of extra spatial dimensions, then, the size of such distances will have to be reconfigured.

Furthermore, while assuming that the two branes are parallel to one another might simplify issues both conceptually and mathematically, the condition is fairly arbitrary, if not completely artificial. There are very few things, if any, in the real world that are parallel ... although treating some of those things as if they were parallel might lead to useful results.

In any event, the two branes are, for the most part, considered to be independent of one another. The only connection the two branes hold in common is the force of gravity since, according to the general theory of relativity, gravity is inherent in spacetime, and, therefore, penetrates to all levels of spatiality – both visible as well as those that might be curled up.

Steinhardt and Turok assume that some kind of an attractive force exists between the two branes. However, in the beginning, the magnitude of the force is fairly small.

Presumably, the magnitude of the force depends, at least in part, on how that force varies as a function of the distance between the two branes. Moreover, the magnitude of that force would also depend on the way it interacts, if it does, with the number of spatial dimensions to which the two branes give expression.

Knowing how the foregoing factors affect the way a force manifests itself might be important when it comes to trying to figure out the minimum distance two branes would have to be from one another to be able to generate the sort of energies that are necessary to fuel a Big Bang. Assuming there are two parallel branes that are separated by the requisite minimal distance and are drawn together in just the right manner seems rather arbitrarily fine-tuned.

Proceeding in the foregoing theoretical manner does not necessarily generate a proof of concept. Rather, what it shows is there are ways of fooling around with qualities and quantities that lead to heuristically valuable results that inevitably lead to questions about whether, or not, reality could have found its way to such an arbitrary arrangement of conditions and properties on its own.

According to Steinhardt and Turok, the two parallel branes also are subject to quantum fluctuations. These fluctuations are amplified by the presence of the attractive force that exists between those two objects.

Over a period of time – perhaps infinite in length – the two branes are drawn toward one another along a fourth dimension. With the passage of time, the strength of the attractive force between the branes increases, eventually resulting in a collision between the two objects.

The kinetic energy of the collision is translated into hot radiation. That radiation is spread throughout the respective branes as a result of the expansion that occurs following the Big Bang.

Due, in part, to the presence of quantum fluctuations, the surfaces of the branes do not collide all at once. They do so over a period of time, and, as a result, the energy of the collision is released in stages.

The expansion that ensues from the Big Bang causes matter and radiation to spread out, and this leads to regions of lower energy density. Dynamic interfaces, on the other hand, involving quantum fluctuations that temporarily pushed surfaces apart would lead to regions of higher energy density because those regions would be subject to the expansive forces associated with the Big Bang that arose through the initial collision for a lesser period of time.

According to Steinhardt and Turok, if the strength of the attractive force between the two branes increased with sufficient rapidity, then, this would generate energy density variations that were capable of exhibiting properties that reflect features of the real world. For example, the variations in energy density observed in the calculations associated with their cyclical model exhibited scale invariance and, therefore, would reflect the same kind of density variance when considered in relation to large-scale structures of the universe (e.g., galaxies, clusters, and superclusters).

In short, if one fed the right sorts of values into their model, and established the right kinds of initial conditions, then, one ended up with calculations that displayed properties similar to what are observed in the universe with which we are familiar. They referred to the product of their model as an ekpyrotic universe because the latter was born in the high, hot, radiation that was generated through the collision of branes.

Wondering whether, or not, branes, generally speaking, have some sort of heuristic value when used to model phenomena outside of the creative imaginations of theorists raises various issues. Wondering whether, or not, such branes: Come in parallel pairs separated by as little as $10^{-30}$ centimeters, interact along a fourth dimension, as well as share an attractive force that flows between them and increases in just the right way involves an array of much more problematic issues.

Furthermore, while assuming that the three dimensions associated with the two branes extend out to infinity might make things easier mathematically, invoking the presence of infinity always makes things harder when it comes to reconciling mathematics with reality. Similar remarks should be made in conjunction with the contention of Steinhardt and Turok that two branes might be drawn toward one another for an infinite length of time prior to the point where the conditions for a Big Bang might arise.

Steinhardt and Turok believe that dark energy might play a central role in their theory. In fact, dark energy (along with gravity) is what drives the cyclical properties of their ekpyrotic universe.

More specifically, dark energy gives expression to one dimension of the spring-like force that ties the branes together. Prior to the Big Bang, dark energy dominates, and, according to Steinhardt and Turok, the two braneworlds would, through the process of expansion, become relatively smooth, flat, and, as well as, exhibit an energy density that was fairly sparse.

However, as the two branes are brought together through the gravitational dimension of the spring-like force that connects them, the potential energy of the branes gets converted into kinetic energy. In turn, a portion of that kinetic energy is transitioned into radiation and particles.

For a time, the matter and radiation that are created through the collision would dominate what was taking place within the respective branes. Eventually, that matter and radiation would be thinned out through the process of expansion and a point would be reached when the potential energy of the brane exceeded the remaining kinetic energy of the matter and radiation that had been generated through the Big Bang.

At this point the spring-like force of dark energy would begin to dominate. At a certain point, the expansive properties of the spring-like connection between the branes would shut down, and the gravitational dimension of that connection eventually would bring about another Big Bang, and a new cycle would begin.

Aside from the critical issues that already have been raised in conjunction with the Steinhardt/Turok model, inserting the idea of dark energy into the discussion introduces a potential source of new problems. As the discussion in 'Chapter 8: Expanding Horizons' indicated, there are reasonable grounds for asking whether, or not, dark energy exists at all, or, if it does exist, whether it exists to the extent that the Type 1-a Supernovae research suggests might be the case.

If dark energy does not exist, or if it does exist but constitutes far less a percentage of the matter/energy of the universe than presently is believed to be the case by those who accept the Type 1-a Supernovae research, then the spring-like mechanism of the ekpyrotic universe might not be sufficiently powerful and flexible to subsidize even one Big Bang, let alone a series of them. Furthermore, while assuming that one brane is heavily shaped by the presence of dark energy entails a variety of problems in and of itself, nonetheless, assuming that there are two branes that are tied together through the spring-like properties of dark energy seems, theoretically speaking, to be pushing things in a fairly arbitrary direction.

According to Steinhardt and Turok, the spring-like connection between the branes keeps the separation between the branes within fairly constant parameters. They indicate that just $10^{-25}$ seconds after the Big Bang collision, the branes would have come to rest at their original positions relative to one another.

However, the branes themselves are able to expand or stretch exponentially. Over a period of a trillion years, or so, the two branes supposedly will double in size a hundred times or more.

When they provided an overview for their theory, Steinhardt and Turok posited a set of starting conditions. One of those conditions indicated that the two branes extended out infinitely in three directions.

Assuming that such spatial extensions are infinite in character permits one to avoid the question of what happens if there are cosmological boundaries that place constraints on how far a given brane can expand or stretch. However the fact of the matter is, we have no idea what the spatial or dimensional limits are, if any, of the universe, and, therefore, permitting branes to extend to infinity is a fairly arbitrary theoretical move.

Currently, the universe is considered by mainstream astronomers to be some 14 billion years old. Yet, Steinhardt and Turok are talking in terms of cycles of a trillion years, or so, during which branes double in size a hundred times.

A trillion years is the time during which the two branes, under the influence of expansion or dark energy, each will become flat and smooth, as well as become parallel with one another. What reason do we have for supposing that the universe is likely to last for a trillion years or be able to double in size a hundred times or more?

Although what Steinhardt and Turok assume might be true, there are no good reasons for supposing that the brane dynamics being described by Steinhardt and Turok constitute a realistic model for the universe in which we actually live. More importantly, short of waiting a trillion years to see what happens, there is no way to empirically substantiate that such theoretical ideas represent realistic conditions vis-à-vis our universe.

While the cyclical model of Steinhardt and Turok does avoid the infinite temperatures and densities of the Standard Cosmological Model (the two theorists talk, instead, of plasma temperatures involving $10^{23}$ degrees), their model doesn't avoid the problem of infinity. Getting rid of density and temperature infinities -- by assuming that: Spatial dimensions are infinite in nature, or that an infinite amount of time might have elapsed before two branes collide to create a Big Bang, or that the Cosmos runs in trillion year cycles that disappear into the infinite future -- seems a rather futile exercise in which one introduces a new set of infinities to replace the set of infinities that one just has jettisoned.

The finite temperatures envisioned by the Steinhardt and Turok model do provide an explanation for why no one has, so far, observed magnetic monopoles … an explanation that is different from the one entailed by the inflationary theory put forth by Guth. According to Guth, the brief period of intense inflation that occurred during the Big Bang diluted the presence of magnetic monopoles to such an extent that they have become too few and too far between to be readily observed, whereas Steinhardt and Turok say that the temperatures associated with their own version of a Big Bang would not have been sufficiently high to enable magnetic monopoles to form.

Finally, Steinhardt and Turok believe their model provides an answer to anyone who might try to argue that the cyclical dimension of that model runs contrary to basic, physical precepts involving the conservation of energy or the tendency of the universe to always increase in entropy or disorder. More specifically, they refer to a long-recognized property of gravity that appears to suggest there is no upper limit to the amount of energy that can be borrowed from gravity.

The two theorists feel that the spring-like force connecting the two branes constitutes a viable means through which their system can continually borrow from the gravitational field that is present in order to subsidize subsequent Big Bangs. Every one trillion years, or so, the energy that is borrowed from the ubiquitous presence of gravity is converted into radiation and matter through the force of colliding branes, setting in motion subsequent stages of their model.

One should note in passing that Steinhardt and Turok do not believe that Einstein's general theory of relativity is up to the task of being able to describe the sort of Big Bang collision they have in mind. Instead, they believe that in order to describe the dynamics of space and time properly, one must utilize theories involving strings and branes.

Whatever the limits of general relativity might be, string theory has not, yet, been shown to constitute a viable candidate for superseding anything. Moreover, if the universe does not give expression to the sort of higher spatial dimensions in which branes reside, then, branes will not be able to show how to establish an account of quantum gravity that would be capable of replacing general relativity theory either.

However, even if one were to grant that, potentially or in principle, there is no upper limit within such a framework with respect to how much energy can be borrowed from the presence of gravity, such a possibility by itself is not sufficient to make the Steinhardt/Turok model work. Branes, strings, and extra spatial dimensions must come together in just the right way to create an ekpyrotic universe, and this set of 'just-so' conditions involves: Two branes existing in close spatial relationship with one another that are regulated by the spring-like force that connects those two parallel objects; branes that are separated from one another by a distance that is set, in part, by the kind of spatial dimensions that are present in those branes; spatial dimensions that extend to infinity; cycles that run for a trillion years and involve expansions that double the size of the branes by a factor of a hundred times; a spring-like force that gives expression to both dark energy and gravity in which, for unknown reasons, the former turns off and on at appropriate times.

Cycles are a function of the entire system. One cannot take advantage of the idea that there is no upper limit to how much energy can be borrowed from gravity (assuming this is true) unless the entire system of which gravity is a part works in the way that Steinhardt and Turok claim is the case. However, as has been indicated at various points during the discussion that has been going on for the last fifteen pages, or so, there are quite a few reasons why one might question the viability of their model.

Moreover, even if there were nothing problematic in the Steinhardt/Turok cyclical model and, therefore, even if one were to accept their premise that there is a way to construct a working theory of cosmology that differs substantially from the Standard Model of Cosmology, none of this necessarily has much to do with the universe in which we find ourselves. In other words, unless, for example, one can show that there are two objects in our universe that operate in the way that the two branes in the Steinhardt/Turok model do, then, their model doesn't necessarily have anything to do with our universe or how our universe got to be the way we observe it to be today.

Chapter 10: Odds and an End

Many theoretical physicists refer to the idea of a "Landscape" in order to try to place physical possibilities in a context that makes sense to them ... and, hopefully, might make sense to other individuals as well. The Landscape to which they are referring is a mathematical construct, and it is a realm that consists of hundreds -- and, possibly, thousands -- of dimensions.

Each entry in the set of all possibilities that are encompassed by the notion of a Landscape gives expression to a vacuum of some kind. In the context of the Landscape, a vacuum involves the set of constants, particles, forces, and laws of interaction that make up the physical background that shapes and constrains what takes place within such a possibility.

Each vacuum has its own set of physical laws. Different laws of physics describe what takes place under different vacuum conditions, and each such vacuum condition is one of the possibilities occupying a place in the Landscape.

The laws of physics that are inherent in the Standard Model of quantum physics are the laws that describe the possible Landscape that we refer to as our universe. Other members of the Landscape would be characterized by laws of physics that differ (in all ways, many ways, or in a few ways) from the laws we use to describe the form of reality that fills the dimensions of our universe.

Each entry in the set of possibilities that make up the Landscape represents an alternative universe. As such, our universe is but one alternative among an indefinitely, if not infinitely, large set of alternative universes, each governed by its own set of physical properties and dynamics.

Any given universe can be represented in terms of a mathematical space. If the objects occupying a universe operate in accordance with the properties of a certain number of fields, then, the mathematical space describing that universe consists of a comparable number of dimensions.

There is, however, a certain amount of ambiguity that surrounds the issue of dimensionality. Some individuals treat dimensions as a mathematical way to spatially represent the variables that are present in a given context, while other individuals believe that each dimension constitutes an actual spatial 'direction' that exists within a given universe.

For example, some individuals maintain that if there are five fields that govern what transpires in a given universe, then, that universe can be represented by a mathematical space consisting of five dimensions. If there are n-fields that underlie the phenomena of a given universe, then the mathematical space that represents such a possibility consists of n-dimensions.

Other individuals proceed in a different manner. They believe that dimensions refer to actual spatial or structural possibilities inherent within a given universe, and, as a result, various fields might interact with those spatial dimensions according to the nature of the different mathematical principles that might regulate such interactions.

Treating dimensions as spatial, structural features of a given universe could lead to a different kind of physics than might arise when dimensionality refers just to the number of variables that have to be taken into consideration when trying to describe the phenomena that take place within such a universe. The foregoing difference might seem to be of a trivial sort, but a great deal of confusion can arise if one is not clear about how the notion of dimensionality is being used to describe a given universe.

Each entry in the Landscape consists of some kind of energy configuration. Such configurations involve: Potential energy, kinetic energy, the laws that govern how the two forms of energy manifest themselves in relation to one another (e.g., conservation laws if any), and, as well, the tendencies that characterize what is likely to transpire over time in such a universe with respect to energy (e.g., Does some sort of principles of entropy operate within such a universe?).

The fields that make up the vacuum condition for any given possible Landscape manifest energies of one kind or another. Those fields might be scalar (quantitative without being directional, such as in the case of temperature) or vectored (quantitative and directional, such as in the case of momentum).

In addition, energies can be in a stable or unstable state. Unstable energies might decay into more stable states (or they might not), and the modalities of decay that are possible might vary from universe to universe within the Landscape.

The Landscape consists of possibilities that involve all manner of combinations involving vacuum states, particles, forces, fields, scalar quantities, vector quantities, forms of energy, dynamics, stabilities, instabilities, and laws that describe the sorts of phenomena that arise when all of the foregoing components interact with one another. As such, the Landscape can be depicted as a series of peaks, hills, valleys, plateaus, crevices, and so on that consist of possibilities involving different arrangements of quantitative values (ranging from maximum to minimum) for an array of states, fields, dimensions, transformations, constants, forces, energies, particles, and physical laws.

As noted earlier, the laws of physics that govern a particular possibility within the Landscape are likely to be different from the laws of physics that govern other possibilities within the Landscape. Nonetheless, while acknowledging the foregoing point, one still could ask about whether, or not, there might be some set of fundamental laws governing the Landscape as a whole that regulate what is and is not physically possible.

There is a potential difference between logical possibility and the possibilities that -- existentially and ontologically speaking – can become manifest as physical realities. Are the possibilities that are logically conceivable constrained in various ways by the actual nature of reality?

Neither theoretical nor experimental physicists are able to answer the foregoing question. The Landscape is a mathematical construct that explores logical possibilities from a mathematical perspective and, as such, is constrained by the limits of what can be imagined mathematically.

String theory projects the possibility of there being at least $10^{500}$ different worlds or alternative universes, each with its own arrangement of conditions giving expression to a vacuum state that forms the heart of such a possibility. To date, theorists have not been able to devise a way to determine if the foregoing $10^{500}$ possibilities contain the sort of vacuum condition that entails the array of constants, energies, forces, particles, fields, dimensions, and laws that are capable of describing the universe in which we find ourselves, and string theory does not necessarily exhaust the possibilities that are encompassed by the Landscape.

As a result, many individuals have questioned the likelihood of ever being able to happen upon some sort of selection principle that would enable one to identify the precise vacuum state from among the indefinitely large number of possibilities that populate the Landscape that would generate the world with which we are familiar. Presumably, such a selection principle would be in the form of some kind of cosmological theory that explains how one makes the transition from first principles involving constants, vacuum state, forces, energies, dimensions, and so on to a universe that, among other things, has the sort of large-scale, dynamic structure of stars, galaxies, clusters, superclusters, and voids that are observed in the universe today.

Although, over the years, a variety of theoretical perspectives have come forth to contend for the championship belt of the cosmos, their game plans all have displayed important theoretical and empirical deficiencies (some of which have been pointed out during the course of this book). Even the reigning champion, the Standard Model of Cosmology that features: a Big Bang, inflation, a cosmological constant, dark matter, and dark energy (the so-called Lambda-CDM – Cold Dark Matter – scenario) seems to be a champion more by virtue of its fan base than because it has been able to prove itself consistently within the context of a fair fight.

Naturally, different commentators call things as they see them. However, there is a concern in some quarters that the people who are judging such contests might be heavily influenced by their own vested interests.

Notwithstanding the foregoing considerations, there are some theorists who believe that the search for a unique selection principle through which to identify which member of the Landscape constitutes our universe is a futile one. Instead, they believe that one should begin to look at the Landscape as being more than just a theoretical construct.

More specifically, they contend that all the possibilities that appear in the Landscape are real. Our universe is just one among many on-going realities that simultaneously occupy different ontological (not theoretical) niches within the array of arrangements of energies, constants, dimensions, and so on that make up the physical topographies that give expression to various existential dimensions of the Landscape ... the set of all such realities or alternative, pocket universes.

The idea that the Landscape, with all its possibilities and potential, constitutes reality is an understandable one. In other words, one has no difficulty in grasping the general features that are entailed by such an idea.

However, claiming that the Landscape has ontological reality does face one small challenge. There doesn't seem to be any evidence to support such a contention.

A given system can be stable, unstable or metastable. Stable systems, whatever their dynamics might be, continue on in a way that preserves the tendencies inherent in those dynamics, while unstable systems contain the seeds (which could be in the form of fields, particles, and/or forces) that lead to the decay, breakdown, or transition of such a system that gives rise to a subsequent system that is stable, unstable, or metastable.

Metastable systems have the capacity to give the appearance of being in a stable state for a long period of time, but, then, something happens -- internally and/or while interacting with other systems – that leads to a sudden, unanticipated transformation in the dynamics of such a system.

Some jugglers give expression to the idea of a metastable system. For extended periods of time, the individual is able to manipulate a number of objects in a way that keeps the necessary number of objects in the air as the right number of objects are handled by her or his hands to keep the dynamic in cyclical, harmonious motion.

Eventually, distraction, tiredness, and/or loss of focus enter into the system. Disaster ensues.

In physics, certain conditions appear to have the potential to give expression to a metastable system. For example, suppose there is some sort of a quantum particle that is situated between several kinds of energy barriers.

What is the probability that such a particle will be able to escape its present position? Calculating such a probability will depend on a number of different variables, but, as long as that figure is not zero, then, presumably, if one waits for a sufficiently long period of time, then for reasons that might not be known, the particle will be able to escape its current state.

The only thing that would legitimately permit one to establish a non-zero probability is if one had observed at least one occasion in which the particle appeared on the other side of the energy barriers that previously had surrounded it. Even then, if the event being observed was a one and done sort of phenomenon, then, the fact that one had observed the particle make such a transition on one occasion carries absolutely no implications for the future likelihood of such an event taking place again.

Quantum tunneling is a term that is used to refer to the foregoing phenomenon. No one really knows what is transpiring when quantum tunneling occurs, but although fairly rare, there is empirical data indicating that such events do occur and have occurred on more than one occasion, and, consequently, such events do not constitute a one and done sort of phenomenon ... meaningful probabilities can be attached to their likelihood in any given set of circumstances.

When various energy barriers surround the aforementioned quantum particle, the system appears to be fairly stable. However, given enough time, the system will break down and the quantum particle will undergo a significant transition.

Apparently, the foregoing quantum particle is subject -- as, supposedly, all quantum objects are – to the vagaries of quantum fluctuations. Nevertheless, quantum fluctuations do not necessarily constitute a description of the actual dynamics that are ontologically taking place with respect to quantum particles, but, instead, quantum fluctuations refer to an array of probabilities that a given particle will, or will not, show up under certain conditions exhibiting various properties.

The potential to show up in more than one place manifesting different kinds of properties does not necessarily mean that a given quantum entity is fluctuating. For example, the fact that dice can show up on different occasions manifesting more than one value does not mean that the dice are fluctuating.

Instead, the probability distributions of quantum mechanics might only indicate that the behavior of quantum entities is shaped by a number of unseen, but determinate, influences and factors. Such factors affect the nature of what is manifested on any given occasion, and such possibilities can be captured through the equations of quantum mechanics.

The foregoing influences and factors cannot be random in character. If they were, then, the quantum equations that generate probability distributions would not lead to reliable results.

The behavior of quantum particles is predictable. The parameters of possibility are determinate.

The fact there are probabilities associated with quantum particles that are of a fairly low order of likelihood does not make such possibilities a function of random events. Rather, it merely means that our ignorance concerning the realities of quantum dynamics is sufficiently great that we don't understand why a quantum entity might behave in one way rather than another in any given instance.

Quantum fluctuations refer to the parameters of possibility that are associated with a small-scale system involving a limited number of particles. Quantum fluctuations do not describe the large-scale potential of the universe.

The notion of quantum fluctuations is not scale-invariant. The methods one uses to describe some of what goes on with a limited number of particles on a quantum level does not describe what is transpiring in the universe as a whole, and this is one of the reasons why theoretical estimates involving the energy density of the vacuum have been off by a factor of $10^{120}$.

Consider the following. One supercools water by taking a certain amount of pure water and, then, very slowly and meticulously, one lowers the temperature of that water below freezing.

When supercooled in the foregoing fashion, adding just a small amount of ice is often sufficient to induce supercooled water to crystallize around the recently introduced ice crystal that has served as a seed of transformation. Nevertheless, until the aforementioned seed of ice is added, the supercooled water can maintain its non-frozen state for an extended period of time while continuing to reside in temperatures that are below freezing.

The foregoing process is referred to as bubble nucleation. The expanding field of crystallizing ice is likened to a bubble that increases in size.

Such a supercooled liquid appears to be stable. However, when the right factor is introduced into such a system, rapid transformations take place as the system undergoes a phase change, and, therefore, the system is really metastable in character.

Some vacuums might be characterized by the foregoing kind of metastability. Such vacuums appear to be stable but under the right conditions, they undergo a phase change – similar to bubble nucleation in supercooled water -- and give expression to unexpected kinds of phenomena.

There is a boundary that separates that part of the old system which has not, yet, undergone the foregoing sorts of phase change from those portions of the system that already have undergone the phase change. This boundary is known as a "domain wall" and, in many ways, it behaves like a membrane that is characterized by its own set of dynamics which mark the advancing front of the phase change that is spreading throughout a previously stable system.

According to some theorists, quantum fluctuations in the Landscape could serve as the seeds for a process that is similar to the previously described dynamic of bubble nucleation. Although many of the foregoing bubbles might come to nothing and just dissipate, some of those bubbles might entail a change in the way such a vacuum manifests itself as the latter undergoes a phase change of some kind relative to previous bubbles and, then develops to more complex states.

The notion of bubble nucleation, the issue of phase change, the condition of metastability, and the existence of various kinds of vacuum conditions are all interesting ideas. However, none of the foregoing ideas – either by itself or in combination with one another – demonstrate that the Landscape exists.

Such ideas allude to possibilities inherent in the potential of the theoretical perspective to which the Landscape notion gives expression. Nonetheless, none of those ideas constitutes the sort of seed that is capable of inducing a phase change or induce a process of bubble nucleation in a metastable Landscape that, thereby, changes a theoretical set of vacuum states into an ontological reality.

If the Landscape is not real, then, it is not subject to either random events or the "quantum fluctuations" that supposedly are described through the method of generating probability distributions that are used in quantum mechanics. In addition, one needs to keep in mind that the Landscape, in its entirety, occupies a scale to which quantum fluctuations (which are appropriate for describing the behavioral properties of small-scale systems) are not necessarily applicable.

Arbitrarily and artificially, one might invest the Landscape with dimensions of quantum fluctuations or random dynamics. However, in doing so, one is merely playing around with theoretical possibilities, and none of that conceptual manipulation demonstrates that the Landscape has an existence beyond the confines of theoretical imagination.

Some theorists envision the Landscape as being capable of subsidizing a process of eternal inflation through which an endless array of alternative vacuum states are formed that develop in accordance with the properties that are inherent in such states. Unfortunately, there is no evidence to indicate that such vacuum states actually are forming, and even if they were forming, there is no evidence to indicate that our own universe came into existence through such a process.

There is no evidence to indicate that a Landscape exists that constitutes a source for an indefinitely large, if not infinitely large, continuum of vacuum states. There is no evidence to indicate that our universe is part of such a continuum.

Even if such a Landscape exists, we have no idea what principles and properties govern its mode of operation. Conceivably, such a Landscape is constrained in some ways, while being enabled in other ways, and, therefore, one cannot automatically assume that the Landscape, even if it were to exist, constitutes a continuum of possibilities that operate in accordance with quantum field theory.

If the Landscape is not infinite in character, then, there is no guarantee that there must be one vacuum state amidst the possibilities encompassed by such a Landscape that will turn out the way in which our universe has turned out. Moreover, even if the Landscape were infinite in character, this does not guarantee that all possibilities will be exhausted through such an infinity any more than the alleged infinity of natural numbers guarantees that irrational and imaginary numbers must exist somewhere in the midst of that collection of natural numbers.

Papers have been written and presentations given that are preoccupied with calculating the rate at which processes akin to bubble nucleation might take place in an inflating universe. Each bubble is given a different cosmological constant with which to work.

The aforementioned perspectives are developed in accordance with agreed upon ideas concerning the nature of quantum field theory. Therefore, such scenarios are considered by many theorists to be quite reliable.

According to the foregoing sorts of exercises, the rate of generating pocket universes might be small, but it is not zero. Be this as it might, an infinite number of monkeys, typing on an infinite number of typewriters supplied with infinite amounts of paper and ribbons, and typing for an infinite amount of time might only generate the same nonsense an infinite number of times and/or produce an infinite number of broken or compromised typewriters ... whichever comes first. One cannot suppose that supposedly random events and/or quantum fluctuations will fare any better than an infinite group of monkeys since there is nothing which guarantees that a set of random events – even if infinite in length – must contain all conceivable possibilities somewhere within its sequence any more than the infinities associated with a group of monkeys and typewriters necessitates that one, or more, classic pieces of literature must, sooner or later, emerge from the ontological mess that is likely to be generated by such a group of monkeys.

Moreover, what if there is not an infinite amount of time available for pocket universes with different vacuum states to bubble into existence? One cannot assume that a finite amount of time, even if indefinitely long, will necessarily and automatically produce a vacuum state that matches that of the universe in which we find ourselves.

In addition, what if not all entries in the Landscape are governed by principles of quantum field theory? Using quantum field to guide one's mode of calculating probabilities seems rather pointless if the possibilities one is making calculations about operate in accordance with some dynamic that is a function of something other than quantum field theory.

What is the probability that a Landscape exists in the sense envisioned by theoretical physicists? The probability is zero.

I'll refer to the foregoing contention as the Whitehouse conjecture. I shall await proof that the conjecture is false.

The countervailing proof for which I am waiting is empirical, not theoretical. I want evidence that the Landscape to which theoretical physicists are referring involves a concrete, tangible, substantive ontological reality that reflects the character of their thoughts concerning the properties of such a reality.

In the Introduction to this volume (as well as in the other volumes of this series), a reference was made to an exchange of ideas that supposedly took place between Napoleon and the physicist, Pierre-Simon Laplace. Napoleon referred to a book on the system of the world that Laplace had written and inquired why the physicist had not mentioned the Creator in that work. Laplace is reported to have said: "I had no need of that hypothesis."

There are several senses in which the same thing could be said in relation to the efforts of cosmologists to account for the nature of the universe. One of the senses being alluded to above involves the possibility that cosmologists are right about virtually everything they have to say concerning the origins and evolution of the universe, while the other sense in which one has no use for the hypotheses of cosmologists is if they are wrong.

What possibly could be meant by the idea that one has no use for the hypotheses of cosmologists even when the latter individuals are right in conjunction with such hypotheses? The answer to the foregoing question is not as preposterous as one might first suppose to be the case.

Let's assume that physicists and astrophysicists are right about the Big Bang, and let's further assume that they are able to resolve all of the outstanding problems that have haunted that scenario for quite some time. For example, let's assume that scientists uncover the details about how the extremely high temperatures associated with a singularity created conditions in which all of the four basic forces were unified and, then, as the universe cooled down, or as other kinds of symmetry-breaking events took place, the different forces precipitated out in the individual forms with which we are familiar today.

Let's assume that the mysteries of an inflationary universe are demystified and that things not only happened in the way envisioned by an appropriately modified version of Alan Guth's initial inflationary hypothesis, but, as well, scientists have discovered what turns the inflationary field on and off. Let's also assume that all of the details of the Higgs mechanisms have been nailed down empirically, and we have come to understand how the Higgs field (and possible multiple companions) turned on during, or shortly after, the Big Bang event.

Let's assume that the Type-1a Supernovae research has been verified many times over as being reliable, and let's assume we have come to understand the nature of dark energy and how it is capable of stretching the fabric of space. Let's assume that we now understand, with great precision, how intrinsic redshift, tired light, and cosmic dust storms that are kicked up by the electromagnetic dynamics taking place in galactic and extragalactic plasmas do not appreciably affect the conclusions that are predicated on Type 1-a Supernovae research.

Let's further suppose that we have come to understand the nature of dark matter, and that we have learned how to properly incorporate such an understanding into our cosmological theories. Let's assume that we now have a fully delineated and proven Lambda-CDM (Cold Dark Matter) model.

While we are in an assuming mood, let's assume that a full theory of quantum gravity has been put together. Let's also assume that someone finally has shown how superstring theory and branes fit into the universe within which we find ourselves.

In addition, let's assume that a much more detailed understanding of how electromagnetic forces operate within plasmas has arisen. As well, let's assume we have learned how to use such understanding to help build better, more complete models of cosmology.

Let's assume that scientists have been able to definitively eliminate all alternative possibilities concerning the significance of Cosmic Microwave Background Radiation. Let's assume that the so-called Axis of Evil involving that data has been explicated, and the Axis is no longer a source of anxiety that keeps some astrophysicists up at night trying to figure out how to reconcile such data with the Standard Model of Cosmology.

Let's assume that black holes have been shown to be more than a theoretical solution to the equations of general relativity. Let's also assume that we now understand: How black holes form; what their internal dynamics are; how their lifecycles operate, and how they help structure the universe.

Let's assume that we now fully understand how galactic clusters and superclusters were able to form in just 14 billion years. Let's further assume that we now know why such clusters, superclusters and voids are arranged in the layered ways that have been observed.

Finally, let's assume that someone has come along and demonstrated that the Whitehouse conjecture has been proven to be false. In other words, let's assume that one, or more, scientists has uncovered empirical proof that the Landscape is an ontological reality and that its properties and dynamics are exactly as theoretical physicists have imagined such properties to be.

Having accomplished all of the foregoing scientific feats, how could anyone possibly claim that he or she has no need of the hypotheses that have all been empirically vindicated? The answer is fairly simple and straightforward.

A perfect cosmological model does nothing to address the most fundamental questions of human existence. Such a model cannot account for the origins of life, or the origins of: intelligence, understanding, consciousness, language, creativity, morality, or curiosity.

A perfect cosmological model cannot account for the origins or existence of the Landscape. Moreover, such a theory cannot resolve issues involving human identity, purpose, or spirituality.

A perfect model of cosmology might be able to answer any number of technical questions concerning the properties, structure, dynamics, and life cycles of an array of galactic and extragalactic phenomena. However, such a model cannot resolve the essential mysteries that enshroud the nature of being human, and, therefore, I really have no need of such hypotheses.

A perfect model of cosmology would be informative. However, it just wouldn't be informative in the way that is needed to reveal the essential nature of existence or my place in the scheme of things.

To be sure, such a perfect model of cosmology could be used as a basis for generating a hermeneutical perspective concerning the possible meanings of existence. However, this would merely return us to a place with which most of us are already familiar – namely, the Final Jeopardy challenge and how best to give expression to the reality problem inherent in that challenge.

A perfect model of cosmology would be able to help constructively shape some facets of any response to the Final Jeopardy challenge. Nonetheless, such a model still would be missing essential ingredients (the origin issues noted previously) and because that model would be missing such elements, I have no need of the many hypotheses that have led to such a model.

The foregoing set of assumptions belongs in a Dr. Pangloss-like environment (and, remember, Voltaire's character was a professor of métaphysico-théologo-cosmolonigologie) where -- cosmologically speaking -- we lived in the best of all possible worlds. Alas, this is not our condition.

Scientists do not know what made the Big Bang possible or even if it actually occurred. If there was an inflationary period in the early universe, no one knows how it came about, or how it came to a halt, or whether, or not, space is susceptible to being inflated.

Physicists do not know whether, or not, the four basic forces were ever unified. Moreover, physicists do not know what the nature of the symmetry-breaking event or events were that led to precipitation of four forces from one underlying unified force (assuming that this is what happened).

Physicists do not know if there is more than one Higgs field. In addition, they do not know what turned that field, or those fields, on.

Although Newton and Einstein had an understanding of how gravity behaved and how such behavior could be modeled, neither Newton nor Einstein understood the nature of gravity itself. Einstein was wrong – or, at least, he was misleading -- when he said that gravity is geometry, and Newton might have been wrong (we don't know) when he claimed that all objects in the universe have a gravitational effect upon one another.

Black holes might exist, but no one yet has definitely proven their existence. Furthermore, even if they do exist, we are completely ignorant about what sorts of dynamics and phenomena might be transpiring within them.

Dark matter might, or might not, exist. Something certainly seems to be affecting the way various stars travel about within galaxies as well as the manner in which galaxies move in relation to one another.

Some people have hypothesized that references to dark matter are unnecessary and might be more adequately explained as a function of the electromagnetic dynamics that take place in galactic and extragalactic plasmas. However, if dark matter does turn out to be something unlike anything scientists have encountered before in the realm of physics, then, what role, if any dark matter has played in a Big Bang, Steady State, or Cyclical cosmological theory is unknown.

The meaning of redshift is subject to some degree of ambiguity, and the precise nature of that degree of ambiguity is difficult to determine. In part, the ambiguity is difficult to resolve because the response of many astrophysicists and astronomers to the idea of inherent redshift is more akin to children than scientists.

If astronomers like Arp and Margaret Burbidge are wrong with respect to the conclusions they have drawn in conjunction with the issue of intrinsic redshift, then scientists should be able to demonstrate – through concrete, empirical data -- the alleged errors of such research. Scientists shouldn't respond by taking away their right to have access to, and use, the instruments found in various astronomical observatories or make the publishing of such research subject to the whims of people with vested theoretical interests.

| Cosmological Frontiers |

The issue of intrinsic redshift has to be resolved. Until scientists understand whether it is a real or an imagined phenomena, and until they understand the extent to which, if any, inherent redshift undermines the idea that redshift necessarily translates into: Distance, recessional velocity and/or the stretching or space, then one is not in a position to reliably interpret the Type 1-a Supernovae research.

Moreover, although astronomers and astrophysicists believe that they have been adjusting their calculations appropriately to factor in the possible effects of cosmic dust and tired light, I'm not entirely convinced this is the case. Data from the Planck space mission indicates there might be more dust in the universe than was assumed to be the case in relation to interpreting the Type 1-a Supernovae research, and, as well, I feel the issue of tired light needs to be rigorously reconsidered in relation to the process of interpreting redshift data.

If dark energy exists, scientists do not currently understand how it works. On the other hand, they have been able to calculate and describe some of its effects.

As is the case in conjunction with inflationary theory, no one knows what the nature of space is and whether, or not, it is stretchable. If space is not susceptible to being stretched, then, obviously, the redshifts associated with dark energy involve a phenomenon of an unknown nature.

Some progress has been made with respect to incorporating plasma research into cosmological modeling. However, by and large, mainstream astronomers and astrophysicists give little thought to the possible contributions that understanding the dynamics of electromagnetic activity within plasmas might be able to make to the field of cosmology.

Finally, if the Big Bang did not take place – and there are good reasons for questioning whether, in fact, it did occur -- then Cosmic Microwave Background Radiation does not mean what it has been interpreted to mean since the mid-1960s. Furthermore, even if Cosmic Microwave Background Radiation is a leftover remnant of the Big Bang, nonetheless, if the so-called 'Axis of Evil' associated with that background data is not a statistical anomaly, then, it constitutes evidence that there might not be scale invariance between the Cosmic

| Cosmological Frontiers |

Microwave Background Radiation and the large-scale structure of the universe.

The idea of scale-invariance is rooted in some rather complex models of probability concerning the nature of the relationship between micro-scale and large-scale structures. The Axis of Evil might be an indication that such models are flawed in certain ways.

If the Big Bang did not take place, then, what are the alternatives? Although ideas like the cyclical cosmological model developed by Steinhardt and Turok are interesting, they also are peppered with a lot of "ifs".

No one knows if branes are a constructive, heuristically valuable way to develop the field of cosmology. No one knows if the universe operates in accordance with the principles of superstring or supergravity models ... and knowing that ten-dimensional superstring theories and eleven-dimensional supergravity models exhibit duality – that is, are equivalent to one another – might be an important thing to know only if it turns out the universe operates in accordance with the principles of superstrings and supergravity.

Given the foregoing uncertain status of cosmology, one comes face to face with another sense in which one might reply that one has no need of such hypotheses. At the present time, the state of things in cosmology is far too unsettled to offer much support in assisting one to be able to constructively and effectively respond to the challenge of Final Jeopardy and the reality problem that is at the heart of that challenge.

When the whole Sputnik crisis began to unfold, I had just crossed over into the teenage years of my life. Part of that unfolding process involved the tremendous emphasis that began to be given to science education in the United States.

To some extent, I was given, and was able to take advantage of, some of the opportunities that ensued from the flight of the Sputnik and the Cold War with the Soviet Union. I went to a high school that only had about 40-50 students, and, yet, because of state and federally funded science initiatives, I was able to participate in a number of programs in science and mathematics that exposed me to ideas that were unlikely to have been part of the regular science program that, at the time, was in place at my high school.

Between my junior and senior years, I was accepted into a summer program funded by the National Science Foundation. Again, I was exposed to the world of science in a way that would not otherwise have been possible in the high school that I attended.

The foregoing experiences kindled an interest in science that has lasted my whole life. However, for a variety of reasons, I didn't pursue science as a career.

Science didn't seem to be able to answer any of the questions that I had concerning the nature of the universe and my role, if any, in it. These were important issues to me, and so, I began to look in other directions.

The truth of the matter is that I probably didn't have the talent to be a good theoretical or experimental scientist. Nonetheless, I did have the capacity to become scientifically literate and, from time to time in my life, pursued this option with considerable intensity.

My interest in science, as has been the case with whatever else I explored, was an attempt to seek the truth of things. Unfortunately, to some extent, I think that science has lost its way when it comes to the issue of truth.

Today, much of science seems shaped by influences that seek to exploit, often problematically, the process of science for purposes of political, monetary, militaristic, economic, and/or technological gain. Themes of control and ego seem to preoccupy many of the activities of educational and commercial institutions that claim to be governed by the principles of science, and, consequently, all too frequently the search for the truth in such institutions seems to be sacrificed on the altar of the vested interests of those who seek to constrain the directions that such a search might take.

On the one hand, scientists often exhibit an incredible degree of insight into the nature of things. On the other hand, those same individuals also often display a considerable lack of foresight into the problematic ramifications that are present in their discoveries.

Pesticides were a solution until Rachel Carson demonstrated that they weren't. Now, people with vested interests are trying to argue that genetically modified organisms are completely safe despite an increasing amount of research indicating that we ought to be far more cautious in relation to GMOs.

Plastics were an amazing discovery. Today, however, in addition to the many other pollution problems that have accompanied that amazing discovery everywhere it travels, there are thousands of square miles of ocean that contain plastics that are breaking down into microstructures that are undermining life in the oceans and, therefore, via the food chain, undermining life on land as well.

Nuclear energy is due to the brilliance of scientists. Unfortunately, such brilliance did not seem to foresee that there really is no safe way to store and treat nuclear waste materials.

Chemists have invented thousands of compounds. Now, with each passing day, new research is coming forth that is uncovering the toxic underbelly of many of those compounds.

The electronic inventions that are used to carry out scientific research, including cosmology, contain many toxic components and, as well, have left a residue of toxicity behind during their manufacture. Scientists have discovered some of the intriguing magnetic properties of rare earth metals and, now, there are tons of those materials being used for military purposes, and, moreover, the production of the electronic equipment that contain such rare earth metals generates a tremendous amount of environmental toxicity.

Billions are being spent on research in cosmology. Meanwhile, millions of children go hungry and live in poverty.

Higher education subsidizes a great deal of cosmological research. In turn, higher education is being subsidized, at least in part, through a process that burdens many young people with a level of debt that will undermine the quality, if not viability, of the rest of their lives.

Much is made of the upside of science. Unfortunately, many people, including scientists, seem to be actively in denial about the downside of science and the problematic nature of the genies scientists keep releasing from the bottle of discovery.

My doctoral dissertation explored a variety of themes and issues that surrounded the process of trying to interpret or understand some given topic. The dissertation was an exercise in the hermeneutics of understanding ... an exercise in trying to determine some of the principles and problems that were inherent in the process of working toward an understanding of understanding.

Four of the seven individuals who were examiners during the oral defense of my dissertation had a strong background in science, including one individual who was a physicist and another individual who was a biophysicist. Different chapters of the dissertation critically engaged issues involving relativity, field theory, quantum dynamics, chaos theory, chronobiology, mathematics, and holography.

Several of the examiners who had a background in science actively pressed me about whether, or not, what I was doing in the dissertation had much relevance to anything of a practical nature or whether, or not, there would be much interest out in the real world with respect to what I was trying to do. I defended myself sufficiently well to get a yes-vote from each of the members of the orals committee, but I also have come to realize that there was a very real point to the questions being asked by some of the members of the examination committee.

For me, going to university (whether for a Bachelors Masters, or Ph.D.) was never primarily about preparing for a career. I hoped, of course, to be able to find work after completing my degree programs, but the fact of the matter was that I worked my way through university, and I always viewed work as a means to an end ... that end being one of having an opportunity to continue searching for the truth of things concerning the nature of reality.

Consequently, when some of the individuals on the doctoral examination committee asked me whether there was anything of a practical nature present in my dissertation or asked me if I thought anyone might be interested in what I was doing, I understood where they were coming from. They were thinking in terms of career, position, and fitting into the institutions of society ... they were thinking in terms of how the world works.

From their perspective, there were questions about the value or practical implications of my work. From my perspective, I was doing the only practical thing a person can do and that is to search for the truth irrespective of whether, or not, other people consider such efforts to be worthwhile ... provided that such a search does not adversely interfere with the lives of others.

Although many individuals might consider science to be the royal road to truth, I am not one of them. I do acknowledge and appreciate – if not marvel over – the many ingenious accomplishments that science and scientists have helped to make possible, but I am somewhat shocked to discover that after a lifetime of reflecting -- in a largely unofficial capacity -- upon the activities and ideas of science, I don't really feel that I have missed all that much by not pursuing science as a career. In fact, I might be at somewhat of an advantage because I do not feel bound – as many scientists appear to be – to try to reduce reality to a set of physical equations.

Science is a great way to learn how to solve certain kinds of problems. Unfortunately, the kinds of problems that scientific methodology finds tractable are, for the most part, of limited value when addressing the Final Jeopardy challenge with respect to trying to figure out the nature of reality ... all dimensions of reality and not just those facets that are tractable to physical methodology.

When scientists begin talking about Landscape theory and the anthropic principle, I sense the same speculative desperation that exists in those who construct theological arguments in order to try to explain away their ignorance concerning the nature of reality. We spend much, if not all, of our days immersed in Sea of Being of which we are largely ignorant.

One should not waste time with the construction of speculative crafts that aren't seaworthy ... even though they might provide one with a false sense of security. We need to learn how to swim through the waters of life with strokes of critical reflection that will not only help keep one afloat but will, as well, help one to swim toward a viable destination – namely, to acquire a feasible way of engaging the Final Jeopardy issue.

During the first three volumes of the series of books that are directed toward engaging the Final Jeopardy challenge (and the present book is *Volume III* in that series), I have critically reflected upon: Medicine, psychopharmacology, biology, evolution, neurobiology, psychology, quantum physics, cosmology, string theory, dark matter, dark energy, gravitational theory, black holes, and more in search of the truth. While all of the foregoing topical areas have provided a certain amount of food for thought, the search continues.

For those who have been paying attention, the aforementioned volumes have been laying down a path, of sorts, that involves a way (not the way) of engaging the Final Jeopardy issue that is capable of leading to constructive and heuristically valuable results. However, in order to discern the presence of that path, one has to look to the interstitial themes that have been woven into the fabric of the critical reflections that run through the aforementioned volumes.

If being, life, faculties, energy, resources, and circumstances permit, there will be three more volumes in this series. Each of those books will be dedicated to a process of critical reflection that engages a variety of additional topics that are relevant to my search for a way of responding, as best I am able and capable of doing, to the Final Jeopardy challenge and its underlying problem concerning the nature of reality.

Bibliography

Articles

Zeeya Mer Ali, 'Gravity Off The Grid', pp. 44-51, *Discover*, March 2012.

Ross D. Andersen, 'An Ear To The Big Bang', pp. 40-47, *Scientific American*, October 2013.

Zvi Bern, Lance J. Dixon and David Kosower, 'Loops, Trees and the Search for New Physics', pp. 34-41, *Scientific American*, May 2012.

Jan Bernauer and Randolf Pohl, 'The Proton Radius Problem', pp. 32-39, *Scientific American*, February 2014.

Leo Blitz, 'The Dark Side of the Milky Way', pp. 36-45, *Scientific American*, October 2011.

Peter Byrne, "The Many Worlds of Hugh Everett", *Scientific American*, October 21, 2008.

David Castlevechhi, 'Is Supersymmetry Dead?' pp. 16-18, *Scientific American*, May 2012.

Timothy Clifton and Pedro G. Ferreira, 'Does Dark Energy Really Exist?', pp. 48-55, *Scientific American*, April 2009.

Stephen J. Crothers, 'COBE and WMAP: Signal Analysis by Fact or Fiction?

David Talbott, 'The Plasma Universe of Hannes Alfvén', pp. 5-10, Edge Science, October-December, 2011.

Tamara Davis, 'Is the Universe Leaking Energy', pp. 38-47, *Scientific American*, July 2010.

Jonathan Feng and Mark Trodden, 'Dark Worlds', pp. 38-45, *Scientific American*, November 2010.

Douglas Finkbeiner, Meng Su, and Dmitry Malyshev, 'Giant Bubbles of the Milky Way', pp. 42-47, *Scientific American*, July 2014.

Tim Folger, 'Second Genesis', pp. 18-24, *Discover Magazine – Extreme Universe*, Winter 2010.

Tim Folger, 'How Can You Be In Two Places At Once?', pp. 56-61, *Discover Magazine – Extreme Universe*, Winter 2010.

Avishay Gal-Yam, 'Super Supernova', pp. 44-49, *Scientific American*, June 2012.

Donald Goldsmith, 'The Far, Far Future of Stars', pp. 32-39, *Scientific American*, March 2012.

Andrew Grant, 'Night Ranger', pp. 32-35, *Discover Magazine – Extreme Universe*, Winter 2010.

Andrew Grant, 'Enter String Man', pp. 70-73, *Discover Magazine – Extreme Universe*, Winter 2010.

Martin Hirsch, Heinrich Pas, and Werner Porod, 'Ghostly Beacons of New Physics', pp. 40-47, *Scientific American*, April 2013.

Ray Jayawardhana, ' Coming Soon: A Supernova Near You', pp. 68-73, *Scientific American*, December 2013.

Meinard Kuhlmann, 'What Is Real?', pp. 40-47, *Scientific American*, August 2013.

Robert Kunzig, 'Sticky Stuff', pp. 50-55, *Discover Magazine – Extreme Universe*, Winter 2010.

Robert Kunzig, 'The Unbearable Lightness of Neutrinos', pp. 62-69, *Discover Magazine – Extreme Universe*, Winter 2010.

Michael D. Lemonick, 'The Dawn of Distant Skies', pp. 40-47, *Scientific American*, July 2013.

Michael D. Lemonick, 'Big Bang', pp. 12-17, *Discover Magazine – Extreme Universe*, Winter 2010.

Noam I. Libeskind, 'Dwarf Galaxies and the Dark Web', pp. 46-51, *Scientific American*, March 2014.

Don Lincoln, 'The Inner Life of Quarks', pp. 36-43, *Scientific American*, November 2012.

Andrei Linde, "The Self-Reproducing Inflationary Universe," *Scientific American*, November 1994.

Joseph Lykken and Maria Spiropulu, 'Supersymmetry and the Crisis in Physics', pp. 34-39, *Scientific American* 2014.

Christopher P. McKay and Victor Parro Garcia, 'How To Search For Life On Mars', pp. 44-49, *Scientific American*, June 2014.

Michael Moyer, 'Is Space Digital', pp. 30-36, *Scientific American*, February 2012.

Steve Nadis, 'First Light', pp. 38-45, *Discover*, April 2014.

F. David Peat, 'Mathematics and the Language of Nature', *Mathematics and Science*, edited by Ronald E. Mickens, World Scientific, 1990.

Michael Riordan, Guido Tonelli and Sau Lan Wu, 'The Higgs At Last', pp. 66-73, Scientific American, October 2012.

Robitaille, Pierre-Marie, 'COBE: A Radiological Analysis', Progress in Physics, 2009, v.4, 17-42

Subir Sachdev, 'Strange and Stringy', pp. 44-51, *Scientific American*, January 2013.

Eric Scerri, 'Cracks in the Periodic Table', pp. 68-73, *Scientific American*, June 2013.

Caleb Scharf, 'The Benevolence of Black Holes', pp. 34-39, *Scientific American,* August 2012.

Steven Stahler, 'The Inner Life of Star Clusters', pp. 44-51, *Scientific American*, March 2013.

Paul J. Steinhardt, 'The Inflation Debate', pp. 36-43, *Scientific American*, April 2011.

Vlatko Vedral, 'Living in a Quantum World', pp. 38-43, *Scientific American*, June 2011.

Hans Christian von Baeyer, 'Quantum Weirdness?' pp. 46-51, *Scientific American*, June 2013.

Carl Zimmer, 'The Surprising Origins of Life's Complexity', pp. 84-89, *Scientific American*, August 2013.

## Books

Lyndon Ashmore, *Big Bang Blasted!:The Story of the Expanding Universe and How It Was Shown to be Wrong*, Book Surge, 2006.

Halton Arp, *Seeing Red: Redshifts, Cosmology and Academic Science*, Apeiron, 1998.

Peter Atkins, *Four Laws: What Drives the Universe*, Oxford University Press, 2007.

Ian G. Barbour, *Myths, Models and Paradigms: A Comparative Study In Science and Religion*, Harper & Row Publishers, 1974.

John D. Barrow, *New Theories of Everything*, Oxford University Press, 2007.

John D. Barrow, *The Constants: From Alpha to Omega – The Numbers That Encode the Deepest Secrets of the Universe*, Random House, 2002.

David Bohm, *Wholeness and the Implicate Order*, Ark Paperbacks, 1983.

Harold I. Brown, *Perception, Theory and Commitment: The New Philosophy of Science*, The University of Chicago, 1977.

Brian Clegg, *Before the Big Bang: The Prehistory of Our Universe*, St. Martin's Press, 2009.

Brian Clegg, *The God Effect: Quantum Entanglement, Science's Strangest Phenomenon*, St. Martin's Press, 2006.

Frank Close, *The Infinity Puzzle: Quantum Field Theory and the Hunt for an Orderly Universe*, Basic Books, 2011.

Frank Close, *Antimatter*, Oxford University Press, 2009.

Brian Cox & Jeff Forshaw, *Why Does $E=mc^2$?*, Da Capo Press, 2009.

Robert P. Crease and Charles Mann, *The Second Creation: Makers of the Revolution in 20th-Century Physics*, Collier Books, 1986.

Paul Davies, *Cosmic Jackpot: Why our Universe Is Just Right For Life*, Houghton Mifflin, 2007.

Guy Deutscher, *Through the Language Glass: Why the World Looks Different in Other Languages*, Metropolitan Books, 2010.

Harald Fritzsch (translated by Gregory Stodolsky), *The Fundamental Constants: A Mystery of Physics*, World Scientific Publishing Company, 2009.

Louisa Gilder, *The Age of Entanglement: When Quantum Physics was Reborn*, Alfred A. Knopf, 2008.

Malcolm Gladwell, *Blink: The Power of Thinking Without Thinking*, Little, Brown and Company, 2005.

Peter Godfrey-Smith, *Theory and Reality: An Introduction to the Philosophy of Science*, University of Chicago Press, 2003.

Rebecca Goldstein, *Incompleteness,* W.W. Norton & Company, 2005.

Nelson Goodman, *Ways of Worldmaking*, Hackett Publishing Company, 1978.

Stephen Hawking, *A Brief History of Time: From the Big Bang to Black Holes*, Bantam Books, 1990.

Nick Herbert, *Quantum Reality: Beyond the New Physics*, Anchor Press/Doubleday, 1985.

John Holland, *Emergence: From Chaos to Order*, Helix Books, 1999.

Dan Hooper, Dark Cosmos: *In Search of Our Universe's Missing Mass and Energy*, Smithsonian Books, 2006.

Stuart Kauffman, *Reinventing the Sacred*, Basic Books, 2008.

Manjit Kumar, *Quantum: Einstein, Bohr, and the Great Debate About the Nature of Reality*, W.W. Norton & Company, 2008.

Leon M. Lederman and Christopher Hill, *Symmetry and the Beautiful Universe*, Prometheus books, 2004.

Lillian R. Lieber, *Infinity: Beyond the Beyond the Beyond*, Paul Dry Books, 2007.

David Lindley, *Uncertainty: Einstein, Heisenberg, Bohr and the Struggle for the Soul of Science*, Doubleday, 2007.

Mario Livio, *Is God a Mathematician?*, Simon & Schuster, 2009.

Lynn Margulis and Dorion Sagan, *Microcosmos: Four Billion Years of Evolution From Our Microbial Ancestors*, Simon & Schuster, 1986.

Thomas O. McGarity and Wendy Wagner, *Bending Science: How Special Interests Corrupt Public Health Research*, Harvard University Press, 2008.

Melanie Mitchell, *Complexity: A Guided Tour*, Oxford University Press, 2009.

Chris Mooney and Sheril Kirshenbaum, *Unscientific America: How Scientific Illiteracy Threatens Our Future*, Basic Books, 2009.

Paul J. Nahin, *The Story of the Square Root of -1: An Imaginary Tale*, Princeton University Press, 1998.

Naomi Oreskes & Erik M. Conway, *Merchants of Doubt: How a Handful of Scientists Obscured the Truth on Issues from Tobacco to Global Warming*, Bloomsbury Press, 2010.

F. David Peat, *Einstein's Moon: Bell's Theorem and the Curious Quest for Quantum Reality*, Contemporary Books, 1990.

Richard Panek, *The 4% Universe: Dark Matter, Dark Energy, and the Race to Discover the Rest of Reality*, Houghton Mifflin Harcourt, 2011.

J.C. Polkinghorne, *The Quantum World*, Penguin Books, 1986.

Alfred S. Posamentier and Ingmar Lehmann, *The (Fabulous) Fibonacci Numbers*, Prometheus Books, 2007.

Helen R. Quinn and Yossi Nir, *The Mystery of the Missing Antimatter*, Princeton University Press, 2008.

Lisa Randall, *Warped Passages: Unraveling The Mysteries of the Universe's Hidden Dimensions*, Harper Perennial, 2005.

Hilton Ratcliffe, *The Static Universe: Exploding the Myth of Cosmic Expansion*, Apeiron 2010.

Hilton Ratcliffe, *The Virtue of Heresy: Confessions of a Dissident Astronomer*, Author House, 2008.

Mark Ronan, *Symmetry Monster: One of the Greatest Quests of Mathematics*, Oxford University Press, 2006.

Ian Sample, *Massive: The Missing Particle that Sparked the Greatest Hunt in Science*, Basic Books, 2010.

Joseph Schild - Editor, *The Big Bang: A Critical Analysis*, Cosmology Science Publishers, 2011.

Donald E. Scott, *The Electric Sky: A Challenge to the Myths of Modern Astronomy*, Mikamar Publishing, 2006.

Lee Smolin, *Three Roads to Quantum Gravity*, Basic Books, 2001.

Lee Smolin, *The Trouble With Physics: The Rise of String Theory, The Fall of a Science, and What Comes Next*, Houghton Mifflin, 2006.

James D. Stein, *Cosmic Numbers: The Numbers That Define Our Universe*, Basic Books, 2011.

Paul J. Steinhardt and Neil Turok, *Endless Universe: Beyond the Big Bang*, Doubleday, 2007.

Ian Stewart, *In Pursuit of the Unknown: 17 Equations That Changed the World*, Profile Books, 2012.

Ian Stewart, *Why Beauty is Truth: A History of Symmetry*, Basic Books, 2007.

Ian Stewart, *Flatterland: Like Flatland, Only More So*, Basic Books, 2001.

Leonard Susskind, *The Black Hole War: My Battle With Stephen Hawking to Make the World Safe for Quantum Mechanics*, Little, Brown and Company, 2008.

Leonard Susskind, *The Cosmic Landscape: String Theory and the Illusion of Intelligent Design*, Back Bay Books, 2006.

Nassim Nicholas Taleb, *The Black Swan: The Impact of the Highly Improbable*, Random House, 2010.

Nassim Nicholas Taleb, *Fooled by Randomness: The Hidden Role of Chance in Life and in the Markets*, Random House, 2004.

Steven Weinberg, *The First Three Minutes: A Modern View of the Origin of the Universe*, Bantam Books, 1979.

David L. Weiner, *Reality Check: What Your Mind Knows, But Isn't Telling You*, Prometheus Books, 2005.

Peter Woit, *Not Even Wrong: The Failure of String Theory and the Search For Unity in Physical Law*, Basic Books, 2006.

www.ingramcontent.com/pod-product-compliance
Lightning Source LLC
Chambersburg PA
CBHW020629220526
45464CB00001B/81